TECHNICAL SKETCHING

For Engineers, Technologists, and Technicians

Second Edition

DALE H. BESTERFIELD, *PH.D., P.E.*
Professor, College of Engineering and
Technology
Southern Illinois University

ROBERT E. O'HAGAN, *M.S. IND. TECH.*
Assistant Professor, College of Engineering
and Technology
Southern Illinois University

PRENTICE HALL, Englewood Cliffs, New Jersey 07632

Library of Congress Cataloging-in-Publication Data

Besterfield, Dale H.
  Technical sketching : for engineers, technologists, and
technicians / Dale H. Besterfield, Robert E. O'Hagan.—2nd ed.
     p. cm.
  Includes index.
  ISBN 0-13-901935-9
  1. Freehand technical sketching. I. O'Hagan, Robert E.
II. Title.
  T359.B47   1990   88-32180
  604.2'4—dc19   CIP

**Editorial/production supervision and
  interior design: Lillian Glennon
Cover design: Wanda Lubelska
Manufacturing buyer: Dave Dickey**

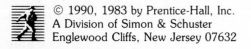 © 1990, 1983 by Prentice-Hall, Inc.
A Division of Simon & Schuster
Englewood Cliffs, New Jersey 07632

Printed in the United States of America

10 9 8 7 6 5 4

ISBN 0-13-901935-9

Prentice-Hall International (UK) Limited, *London*
Prentice-Hall of Australia Pty. Limited, *Sydney*
Prentice-Hall Canada Inc., *Toronto*
Prentice-Hall Hispanoamericana, S.A., *Mexico*
Prentice-Hall of India Private Limited, *New Delhi*
Prentice-Hall of Japan, Inc., *Tokyo*
Simon & Schuster Asia Pte. Ltd., *Singapore*
Editora Prentice-Hall do Brasil, Ltda., *Rio de Janeiro*

To our wives for their
understanding and patience

CONTENTS

PREFACE

During the past decade there has been a significant change in the method of communicating technical information in graphical form. A substantial portion of engineers, technologists, and technicians now utilize sketches to communicate their ideas. Computer-aided drafting and design is taking the place of instrument drawings, and the T-square, triangle, and compass will become as obsolete as the slide rule.

Even though the computer will replace instrument drawings, technical sketching will still be used to generate initial ideas on paper. A knowledge of technical sketching is a prerequisite to learning computer-aided drafting and design and is essential to interpreting working drawing generated by a computer or by drafting instruments.

This book has been designed to enable the reader to:

1. Communicate, by means of a three-view or pictorial sketch, an idea or concept. The sketch should be of sufficient detail, clearness, and accuracy to enable working drawings to be produced by a computer or by instruments.

2. Read and interpret a working drawing.

3. Understand the basic principles of graphic communication as an initial step toward acquiring proficiency in computer-aided drafting and design.

Sketches can be produced in approximately one-sixth the amount of time as instrument drawings; therefore, learning is much more efficient.

Technical programs in community colleges and technical institutes, such as machine tool, automotive, electronics and electrical, maintenance, and welding, to name a few, will find this book an excellent one to meet their needs. Most engineering technology and industrial technology programs in colleges and universities will find that the sketching approach will also meet their needs. The book is ideal for a freshman course or part thereof in an engineering program. It will also serve the instructional needs of in-house programs in industry either on a group or self-instruction basis.

The book begins with chapters on sketching techniques and lettering. This is followed by a detailed description of orthographic projection, pictorials, auxiliary views, and sectioning. The final chapters cover dimensioning, tolerancing, fastening techniques, and working drawing. All the essential areas that are covered in a textbook on instrument drawing are included. Sufficient exercises are given to enable the reader to become proficient in the technique.

The second edition has been improved by updating the dimensioning and tolerancing material to conform to the new standards. A new chapter on computer-aided drafting has been added and all chapters have been updated. Metric units are used throughout. Additional exercises have been included where appropriate.

The authors are grateful to the numerous students and faculty throughout the world who have helped to eradicate errors and clarify explanations.

Dale H. Besterfield

Robert E. O'Hagan

Carbondale, Illinois

SKETCHING

chapter

1

Upon completion of this chapter, the student is expected to:

- **Know the purpose of sketching.**

- **Know the suggestions for good technical sketching procedures and habits.**

- **Be able to construct an acceptable free-hand sketch.**

INTRODUCTION

The importance of technical sketching in engineering and technology cannot be overemphasized. Many original design ideas are conceived by means of sketches. Engineers turn sketches of their ideas over to drafters to construct instrument or computer-generated drawings or to technicians to construct the actual object.

Technical sketching is used as a means of rapid graphical communication that enables the engineer to quickly express, explain, and record her or his ideas. Sketching is done freehand. The time required to make a sketch should be only a fraction (one-sixth to one-tenth) of the time required to make an instrument drawing. All sketches are made with dark, distinct lines and are suitable for reproduction by photocopy or ozalid reproduction.

The principal difference between a technical sketch and a technical drawing is one of degree. Sketches are made freehand, whereas drawings are made with instruments or by the computer.

SKETCHING MATERIALS

One advantage of technical sketching is that it only requires pencil and paper. The types of pencils are grouped into hard, medium, and soft leads. Hard leads make a light thin line, soft leads make a dark wide line, and medium leads produce a line between these two extremes. Numerals and letters are used to designate the type of lead. An HB-grade lead (medium) is recommended to sketch lines that describe the shape of an object and also to letter. A 3H-grade lead (hard) is used to sketch lines that aid in the construction of the object. The medium-grade leads are sufficient for most sketching applications. Lead grades in the medium category are 3H, 2H, H, F, HB, and B, with the B grade being softer than the H grades. The texture of the paper and the humidity may require a slightly different grade of lead. Experimentation is recommended to find the grade that provides the best results. Types of lines are discussed at the appropriate point in the text.

The pencil should be kept sharp at all times. Mechanical pencils with thin leads are available that eliminate the need to sharpen the lead.

Due to recent advances in fiber pens, practitioners have found it convenient to construct an initial sketch with penciled lines, and then to produce dark distinct lines using the appropriate fiber pen. Many designers use various templates to construct circles, ellipses, and various symbols of fluid power, welding, electrical system parts, and so forth.

The other sketching material is paper. Essentially, there are four types of sketching paper: cross section, isometric, transparent, and plain white. They are illustrated in Figure 1-1.

Cross-section or grid paper is the most common sketching paper. Horizontal and vertical lines can be easily sketched by following the grid lines. Different grid spaces, such as four, five, eight, and ten squares to the inch, are available. Special cross-section paper using blue grid lines that do not reproduce by photocopy or other reproduction methods can be obtained. This special paper has the advantage of the grid lines for sketching and a reproduced copy that does not have the grid lines.

Isometric paper is used for a particular type of pictorial sketch. The three different grid-line directions correspond to the three axes of the isometric pictorial. This type of paper is also available with grid lines that do not reproduce.

Transparent paper, which is sometimes referred to as onion skin, tracing, or vellum, has the advantage that light passes through it. Sketch pads of transparent paper are available with various types of master grids. When the master grid is inserted beneath the transparent paper, the grid lines are visible. The sketch is constructed using these grid lines; however, the finished sketch does not show the grid lines. Plain white paper can be used for sketching; however, the other types are easier to use.

Cross section   Isometric

Transparent   Plain white

**Figure 1-1
Types of paper.**

SKETCHING TECHNIQUES

The pencil should be held firmly in the hand with the forefinger extended along the pencil as shown in Figure 1-2. Avoid gripping the pencil tightly with the fingertips, as this is very tiresome on the fingers. The forearm should be rested on a flat tabletop and the motion of the hand should be a wrist motion when making short- and medium-length lines.

**Figure 1-2
Holding the pencil.**

3

(a) Horizontal   (b) Vertical

Figure 1-3
Sketching lines.

Keep eye on
endpoint

Starting point

Figure 1-4
Line between two points.

Avoid sketching positions that do not rest the hand and arm on the tabletop. Grip the pencil at a normal length from the end, approximately as shown in the illustration. It is not necessary to fasten the paper to the table.

The technique for sketching horizontal lines is shown in Figure 1-3a. It requires a wrist motion for medium or short lines, not a whole arm motion. Lines should be drawn from left to right for right-handed people and right to left for left-handed people.

For vertical lines, it is considered best to draw the vertical lines with a downward motion, as shown in Figure 1-3b. This depends somewhat on the ability and preference of the individual person. Many people find it convenient to turn the paper to a position that is comfortable. The basic rule for sketching is to pull the pencil rather than push it against the paper.

One of the best techniques for sketching a line between two points is to start with the pencil at one point and draw the line quickly to the second point, keeping your eye on the second point, as shown by Figure 1-4. It is just as easy to make a solid continuous line as to make a series of short broken lines. Practice this technique and you will achieve excellent results.

Straight lines are the most common type of geometric construction. The circle is the second most common, and the two types account for over 95% of the geometric constructions. The recommended practice for sketching circles is shown in Figure 1-5 and consists of four steps:

Step I. Draw the lines that indicate the center of the circle.

Step II. Lay out the radii on each of these lines, thereby locating four points on the circumference of the circle.

Step III. Locate approximately two points in each quadrant equal to the radius of the circle.

Step IV. Draw in the solid line representing the circumference of the circle.

Many practitioners have found it more convenient to use a circle template, which is faster and neater than the foregoing method.

A three-viewed sketch of an object, using cross-section paper as an aid, is illustrated in Figure 1-6. Notice that the sketch lines stand out above the grid lines. Since the sketch is sometimes used in place of an instrument drawing, it must be complete in every detail, including notes and dimensions.

Three types of lines are shown in Figure 1-6.

Visible or object lines describe the visible surfaces in the various views of the object. They are thick dark lines and are constructed with an HB-grade lead. Visible lines are always the widest and most prominent lines of a sketch.

Hidden lines describe the hidden surfaces in the various views of an object. They are dark thin lines constructed by alternating a short dash [about 3 mm (0.12 inch) long] and a space [1 mm (0.04 inch)]. An HB-grade lead can be used provided that the line is thin. If this is not possible, an H grade may be more appropriate.

Centerlines locate the center of a circle or an axis of a cylinder. They are constructed by alternating a short dash [about 3 mm (0.12 inch) long], a space [about 1 mm (0.04 inch)], and a long dash between 20 mm (0.80 inch) and 40 mm (1.58 inches) long. The center of the circle or of an arc is located by the crossing of two short dashes. An HB-

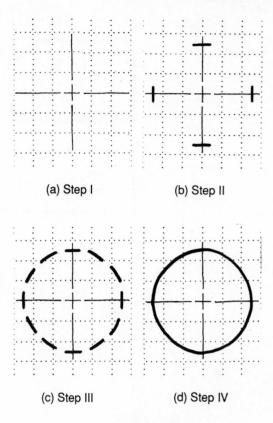

(a) Step I   (b) Step II

(c) Step III   (d) Step IV

**Figure 1-5
Method of sketching circles.**

Figure 1-6
Three-view sketch of an object.

grade lead can be used provided that the line is thin. If this is not possible, an H grade may be more appropriate. Centerlines are shown in all views of an object.

Since objects are, for the most part, made up of regular surfaces, it is not necessary to be an artist to make an acceptable sketch. With practice, sketched lines can be constructed with almost the same appearance as lines drawn with instruments.

PROPORTION

Although a sketch is not made to scale, it must be in the proper proportion. An object may be sketched actual or full size; however, if the object is quite large, it will need to be sketched smaller. Or if the object is quite small, it will need to be sketched much larger than its actual size. The proportion used depends on the relative size of the object and the size of the paper on which it is sketched. Figure 1-7 illustrates a half-size, an actual- or full-size, and a double-size sketch of a steel washer. Of course, other proportions can be used.

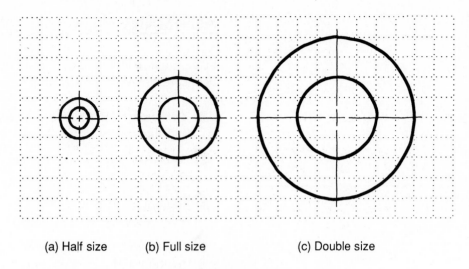

(a) Half size   (b) Full size   (c) Double size

**Figure 1-7
Proportion.**

In sketching an object such as an automobile or some irregularly shaped part, it is often a decided help to use the grid system. First the main controlling points or proportionate size of the object is established on the grid. In this case, Figure 1-8, the bumper-to-bumper length of the automobile, the wheel base, and the height are established. Then the detail parts are sketched. The use of the grid system enables us to keep the parts of the automobile in their proper proportion.

**Figure 1-8
Use of grid lines to obtain proportion.**

SUMMARY

In summary, let's review the information that has been presented.

1. Use an HB-grade lead or fiber pen for sketching the shape of the object.

2. Sketches quickly convey, explain, and record ideas.

3. Technical sketches are not done to scale; however, they are made in the proper proportion.

4. The time required to make a sketch is only a fraction of the time (one-sixth to one-tenth) required to make an instrument drawing.

5. Lines on sketches should be dark and distinct and of the correct type.

6. Paper is not fastened to the table while sketching.

7. When possible, sketch lines should be made in one continuous stroke.

8. Visible lines represent visible surfaces; hidden lines represent hidden surfaces; and centerlines locate the center of a circle or an axis of a cylinder.

9. Cross-section paper and isometric paper are designed to aid in sketching.

10. Sketches must be complete in every detail, as they are sometimes used in place of instrument drawings.

11. Sketches should be of such quality that they can be reproduced.

12. The pencil should be pulled rather than pushed.

13. It is not necessary to be an artist to make an acceptable sketch; however, practice will develop improved proficiency.

LETTERING

chapter

2

Upon completion of this chapter, the student is expected to:

- Be able to convey, on a sketch, verbal and numerical information that is legible, neat, and well proportioned.

ABCDEFGHIJKLMNOPQRS
TUVWXYZ 0123456789

Figure 2-1
Vertical letter alphabet.

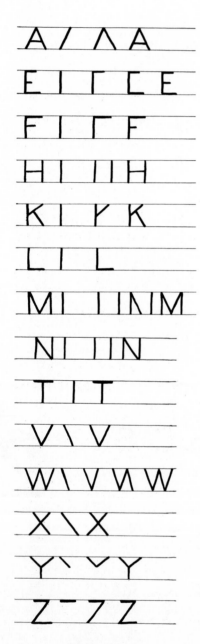

Figure 2-2
Order of straight-line strokes.

INTRODUCTION

This unit deals with the fundamentals and techniques of good lettering. It is a requirement on all sketches that the lettering be legible, neat, and well proportioned. The proper pencil must be used in lettering, and the character and proportions of the letters must be within reasonable standards. This chapter emphasizes vertical lettering, guidelines, and fractions. Inclined lettering and lowercase lettering are not included.

Guidelines are necessary to achieve uniform lettering. The square grid of cross-section paper can be used for guidelines. Lettering is accomplished using an HB-grade lead. With a reasonable amount of conscientious practice, good lettering can be achieved.

VERTICAL LETTER ALPHABET

Both capital letters and numerals for vertical-style lettering are shown in Figure 2-1. The reader should study the proportion of the various letters—width to the height. We can see from this illustration that the letter **I** is the narrowest, while **W** is the widest. A good approach is to separate the letters and numerals into their common groups: the straight-line group and the curved-line group. The two strokes to master are the straight-line stroke and the oval stroke. Once this is accomplished, we are able to form most of the letters either as single strokes or as combinations of straight-line and oval strokes.

The order of strokes used to form the various straight-line letters is shown in successive steps in Figure 2-2. The letter is shown at the left and successively to the right are the steps taken in forming this letter. It is well to analyze these one by one for the proper execution of the letter by the orderly succession of strokes. Strokes are made from top to bottom and left to right so that the pencil is pulled rather than pushed.

The oval-group letters and the successive steps in forming the letters are shown in Figure 2-3. Each letter should be analyzed for the character of the letter, the proportions, and orderly step-by-step execution of forming the letter.

The numerals, as shown in Figure 2-4, should be analyzed one by one and the proper order of strokes studied and practiced. The height of the numerals is the same as that of the letters. Where fractions are shown, the whole number of the fraction is the same height as the letters.

Stability or pleasing proportion of lettering is an important part of the letter or numeral. The top portion of **3, 8,** and **B** should be a little smaller than the bottom to give the appearance of stability.

The need to follow the proper order of strokes cannot be overemphasized.

GUIDELINES

Guidelines are a must to obtain legible, neat, and well-proportioned letters and numerals. For sketching, the common practice is to utilize the square grid of cross-section paper for the guidelines, thereby eliminating the need for their construction. A square

Figure 2-3
Curved-line lettering strokes.

Figure 2-4
Numerical strokes.

GUIDELINES ARE NECESSARY
7.8MM TOP TO BOTTOM

Figure 2-5
Use of the square grid for guidelines.

grid of 5 mm (0.20 inches) is usually satisfactory and is shown in Figure 2-5.

If a square grid is not available, guidelines must be constructed using a straightedge and 6H-grade lead.

SPACING

To obtain a pleasing appearance, the spacing between words and letters is critical. Spacing between letters is a function of the area between the letters, not the distance between the letters. The spacing between words should be approximately the width of the capital letter **O,** as shown in Figure 2-6. The correct (upper line) and incorrect (lower line) spacing between letters is also shown. To make the letters and the space between letters exceptionally

PRACTICE MAKES PERFECT

PRACTICE MAKES PERFECT

Figure 2-6
Lettering examples.

wide is poor practice in that they take up too much space. To make the letters and the space between letters too narrow makes them too hard to read and they tend to blur in reproduction. Attempts should be made to keep the areas between letters and the width of letters uniform and approximately as shown in the figure.

SUMMARY

In summary, one must know or do the following:

1. Know the character and proportions of the letter or numeral to be made.

2. If a square grid is not available, use a 6H-grade lead for guidelines. Use an HB-grade lead for lettering.

3. Use proper spacing between letters and proper spacing between words.

4. Follow the recommended order of strokes for forming the letter or numeral.

5. Strokes are made top to bottom and left to right.

6. Practice sufficiently to acquire reasonable proficiency.

ORTHOGRAPHIC PROJECTION

Upon completion of this chapter, the student is expected to:

- Know the principles of orthographic projection.

- Know the rules for selecting the front view.

- Be able to identify the different types of surfaces.

- Be able to sketch three views of an object from a pictorial sketch.

- Be able to sketch a principal view of an object, given the other two views.

chapter

3

INTRODUCTION

Orthographic projection is a graphical means of communication used by engineers, technologists, and technicians. Special training is necessary in the use of this means of communication. It is a graphical language, and because of the basic principles used in developing the language, it is interpreted and understood by people of different nationalities with different written and spoken languages.

An understanding of orthographic projection involves a study of the basic rules for developing the views of an object, coupled with analysis, visualization, and reasoning. Neither making nor interpreting the orthographic views of an object is a mechanical process.

Orthographic projection is used to create, design, describe, manufacture, fabricate and assemble, repair and replace parts, machines, and concepts. Its origin dates back to Biblical times, and future use is assured. Our industrial, technical, and scientific society is built on the ability to communicate by orthographic projection. Orthographic projection is also referred to as multiview drawings or sketches. The orthographic concept is used in sketching, instrument drawings, and computer graphics.

TYPES OF PROJECTIONS

Sketches of the different types of projections are shown in Figure 3-1. All these projections are of the same object. The most commonly used projection is the orthographic.

The other three projections represent the object in three-dimensional form. *Axonometric* projection has a definite relationship of the three axes and the length along each axis. There are three types of axonometric projection. The most common one is called isometric and it is discussed in Chapter 4.

Oblique projection is actually not a true projection. For some objects it will convey the same information as an axonometric projection and is easier to construct. Oblique sketches are also discussed in Chapter 4.

The *perspective* type of projection is similar to a photograph; however, it is difficult to construct. Its principal use is in architecture.

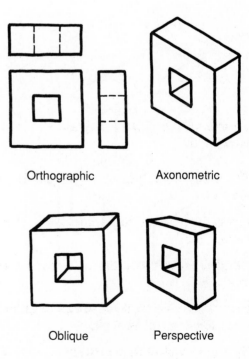

Orthographic   Axonometric

Oblique   Perspective

Figure 3-1
Types of projections.

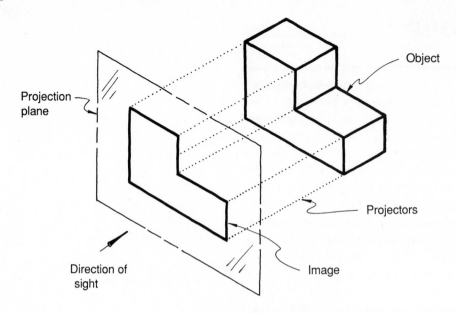

Figure 3-2
Projection of an object.

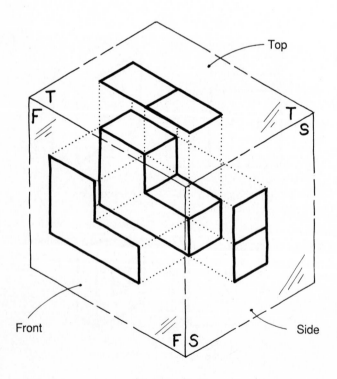

Figure 3-3
Projection box.

PRINCIPLE OF ORTHOGRAPHIC PROJECTION

The fundamental principle of orthographic projection involves the projection of an image of the object on an imaginary plane by use of projectors that are perpendicular to that plane. For this to be theoretically possible, the observer is considered to be at an infinite location. The plane is called a projection plane. This concept is illustrated in Figure 3-2.

One of the best ways to understand the concept of orthographic projection is by means of an imaginary glass box or projection box, as shown in Figure 3-3. The object is inside the projection box and the image of the object is projected to the sides of the box, which are the projection planes. Visible projection planes are the front, top, and side. The six sides of this imaginary glass box are the projection planes, and they are at right angles to each other. Orthographic projection means projection in which the projection lines are perpendicular (at right angles) to the projection plane. *Ortho* means at right angles, and *graphic* means describing in full detail.

There are two systems of projection that are recognized internationally: first-angle projection and third-angle projection. In third-angle projection the object is behind the projection planes as described in the preceding two paragraphs. Third-angle projection is the system used in the United States and Canada. In first-angle projection the object is in front of the projection planes. It is the system used in the rest of the world.

Since it is required to show views of a solid or three-dimensional object on a flat sheet of paper, it is necessary to unfold the six projection planes so that they all lie in a two-dimensional plane such as this page. The six sides of the glass box form the six primary planes of projection. Front, top, right-side, bottom, left-side, and back projection planes are shown in their unfolded state in Figure 3-4.

The image of the object is projected to the planes as shown in Figure 3-5. While there are six primary views, pairs of views are the same. The front and back are the same, except that they are mirror images. The top and bottom views are the same, except that in the top view the vertical surface in the center is represented by a visible line and the bottom view by a hidden line. The right-side and left-side views are the same, except that in the right-

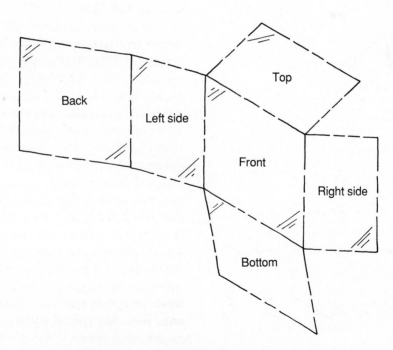

Figure 3-4
The six primary planes of projection.

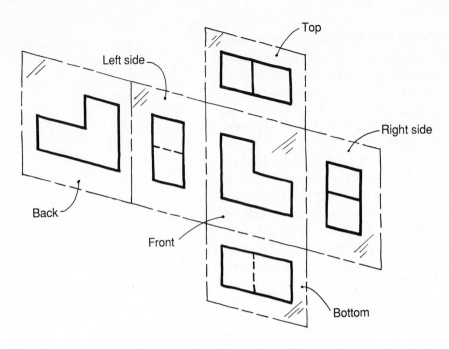

Figure 3-5
The six primary planes of projection
and the object.

side view the horizontal surface of the object is represented by a visible line, and in the left-side view by a hidden line.

Since pairs of views are the same except for visible and hidden surfaces, three projections will usually describe the object. These projections are the front, top, and right side and are called the standard views. In the remainder of the book the right-side view will be referred to as the side view. The front view is the principal view.

Our discussion of the concept continues with a different object enclosed by the imaginary glass box. Figure 3-6 shows an object with normal, cylindrical, and inclined surfaces. The hole is shown, using hidden or dashed lines in two of the views. Centerlines are used to describe the location of the center of the hole and the axis of the cylinder. The projection planes with the image of the object are partially unfolded, as shown in the illustration at the right. Only three views are needed to describe the object.

Note the use of folding lines that correspond to the hinge lines of the glass projection box. Between the top projection plane and the front projection plane is the folding line labeled T/F. Between the front projection plane and the side projection plane is the folding line labeled F/S. Between the top projection plane and the side projection plane is the folding line labeled T/S. In the unfolded position the T/S folding line has two positions.

These folding lines are the edges of the projection planes. For example, the folding line T/F is the edge view of the front projection plane when viewed from the top and the edge view of the top projection plane when viewed from the front. In each view, two of the projection planes appear as an edge.

When the projection planes are unfolded 90°, the result is three views on the two-dimensional page as shown in Figure 3-7. The folding lines are reference lines, which are used in the construction of the three-view sketch. They are erased after the sketch is completed.

There is a definite relationship between the views shown by the lightly drawn lines with the arrows. The width of the object is shown in the front and top views, the height in the front and side views, and the depth in the top and side views. Each feature of the object, such as a hole, is shown in each view. Also each feature is the correct distance from the projection planes as represented by the folding lines.

Figure 3-6
Projection box and three planes unfolded.

Figure 3-7
Orthographic or multiview sketch.

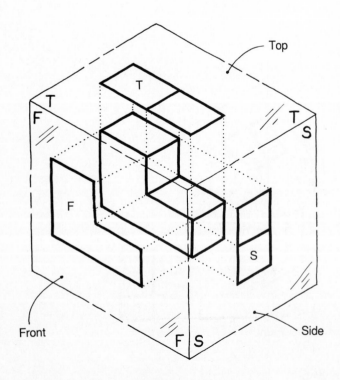

Figure 3-8
Projection box and object.

VIEWING AN OBJECT

The projection box and our original object are shown in Figure 3-8. Also shown are the orthographic views. We will now develop the principles of projection for each view. If you have understood the previous material, this information will be a review. If you are still a little hazy, this section will clarify your thinking on the principle of orthographic projection.

A pictorial sketch and a three-view orthographic sketch are illustrated in Figure 3-9. The image of the object is projected to the front projection plane by level lines of sight in space. This image, called the front view, is sketched by the observer, who is assumed to be at an infinite distance. The top and side views are shown to indicate their relative position in the three-view sketch.

The observer now moves to the top of the projection box to obtain the top view as shown in Figure 3-10. The image of the object is projected to the top projection plane by vertical lines of sight in space. This image, called the top view, is sketched by the observer, who is assumed to be at an infinite dis-

**Figure 3-9
Front view.**

tance. The front and side views are also shown to indicate their relative position in the three-view sketch.

The observer moves to the side of the projection box as shown in Figure 3-11 to obtain the side view. The image of the object is projected to the side projection plane by level lines of sight in space. This image, called the side view, is sketched by the observer, who is assumed to be at an infinite distance. The front and top views are also shown to

**Figure 3-10
Top view.**

Side
projection
plane

Direction of
sight

Figure 3-11
Side view.

indicate their relative position in the three-view sketch.

It is important to remember that the object inside the projection box is left stationary. The object is never rotated; the observer moves to each position.

REVIEW

At this point we can review the principles and basic rules for making orthographic sketches.

1. Orthographic projection means right-angle projection.

2. Orthographic projection uses projection planes.

3. Each set of two adjacent projection planes is perpendicular to each other.

4. The lines of sight (projectors) are parallel to each other and perpendicular to the projection plane.

5. The observer is assumed to be at an infinite distance from the projection plane.

6. A projection is the image of the object on the projection plane.

7. The three most commonly used projections are front view, top view, and side view.

8. The front view is the principal view.

9. Only one view of an object may be seen at one time.

10. Three views are usually necessary to describe an object; however, some objects might only need two or one.

11. Two of the planes of projection appear as an edge in each orthographic view.

12. Folding lines represent the edges or intersections of projection planes.

13. The object is stationary; it is never rotated.

14. A three-view sketch means that the observer has viewed the object from three different directions.

SELECTING THE BEST CHOICE OF THE FRONT VIEW

The person making the sketch selects the best choice for the front view. The object is placed in a stable position and turned so that the characteristic shape shows and unnecessary hidden surfaces are avoided. Figure 3-12 illustrates the best choice of the front view for the L-shaped object.

A poor choice of selecting the front view is shown in Figure 3-13. The front view does not give the characteristic shape of the object; it is the side view that shows the L-shape of the object.

Figure 3-14 shows a different choice for the front view. It, too, is a poor choice. Although we have the characteristic shape shown in the front view, we have introduced an unnecessary hidden line in the side view.

Figure 3-15 also shows a poor choice of a front view. The object is in an unstable position.

The best choice for the front view for a different object is shown in Figure 3-16. This front view shows the object to be rectangular in shape and has a circular hole running through the object. This is the proper choice for the front view of such an object.

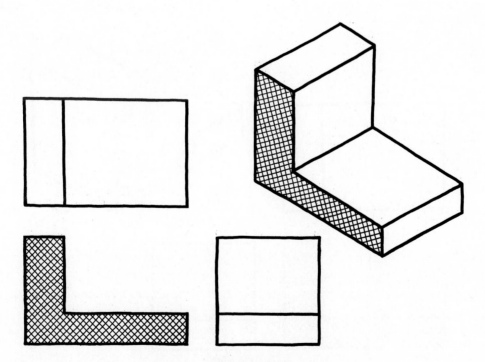

Figure 3-12
Best choice of front view.

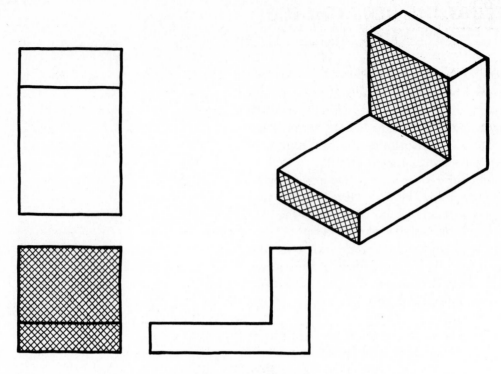

Figure 3-13
Poor choice of front view.

Figure 3-14
Poor choice of front view.

Figure 3-15
Poor choice of front view.

Figure 3-16
Best choice of front view.

The selection of the front view is very important since the proper choice makes the three-view sketch easier to read. Rules for selecting the front view are as follows:

1. The front projection is the principal view.

2. Choose the front view to show the most characteristic shape of the object.

3. The object should be sketched in a stable position.

4. The object should be sketched to see as many visible lines as possible in the front view.

5. If only two views are needed, the front view with either the top or side view is used.

PROJECTION OF A POINT

Since numerous points make up a line or surface, the projection of a point is fundamental to understanding three-view sketches. A point, 1, is inside the projection box as shown in Figure 3-17. Projections of the point are labeled 1 in the front, top, and side views. The folding line represents the plane as an edge. The top plane (folding line T/F) and the

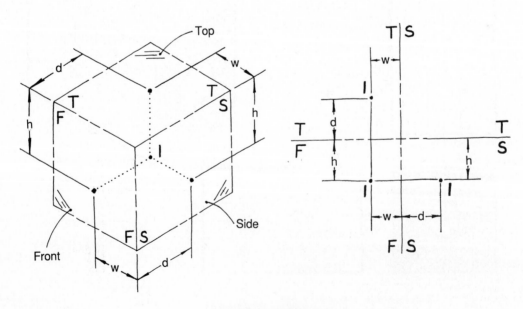

Figure 3-17
Projection of a point.

side plane (folding line F/S) appear as edges when the object is viewed from the front. Likewise, when the object is viewed from the top, the side plane (folding line T/S) and the front plane (folding line T/F) appear as edges. Furthermore, when the object is viewed from the side, the top plane (folding line T/S) and the front plane (folding line F/S) appear as edges. Folding line T/S appears in both the top view and the side view. It is very important to realize that in any given view you are looking perpendicular to one plane and seeing the two adjacent planes as edges.

There are certain common distances in the related views. When looking down on point 1, it will appear at distance d (depth) back of the front plane (folding line T/F). This same distance back of the front plane (folding line F/S), distance d will show in the side view and may be located that way. Notice that the height of the point, distance h (height), is the same in the front and side views. In other words, they are on a level line, parallel to the folding line, drawn across from the front to the side view. The distance point 1 from the side plane is given by the letter w (width). Emphasis should be placed upon which planes appear as an edge in which of the views and where the distance can be measured in respect to that plane.

The folding lines actually become the separation lines for the views. They form a cross whereby the vertical line is the folding line T/S at the top and folding line F/S at the bottom. The horizontal line is the folding line T/F at the left and folding line T/S at the right.

The reader should study Figure 3-17 closely, as the projection of a point is the foundation of orthographic projection. Lines and surfaces are merely projections of points.

Usually, the point in the front view is located first, and then the top view d. The point in the top view is located a distance d from folding line T/F and is directly above and directly in line with the front view, as shown in Figure 3-18a. The side view of the point is drawn last and is in a fixed relationship to the other views. There are three ways of locating the side projection. It may be accomplished with a 45° projection line, as shown in Figure 3-18b and c. This is the most common method of projecting points to the side view. It may be accomplished by a radius, as shown in Figure 3-18d. What we are doing is merely transferring the distance from the

(a)

(b)

(c)

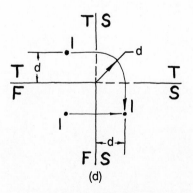

(d)

**Figure 3-18
Locating the side projection.**

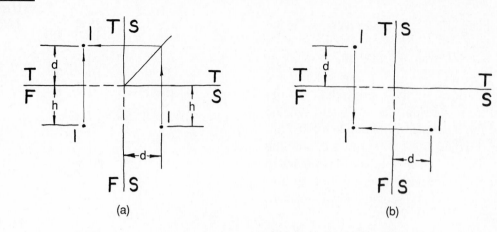

Figure 3-19
Locating the top and front projection.

front projection plane as it appears in the top view to the same distance from the front projection plane as it appears in the side view, and this could be accomplished using just the grid lines. Remember that the other coordinate of the point in the side view is obtained by sketching a level line, parallel to the folding line, from 1. In complicated sketches, it is frequently necessary to label points to obtain the side view. This practice will be discussed later in the chapter.

If the point is located first in the front view and then in the side view, the same technique can be used as shown in Figure 3-19a. The point in the side view is located a distance (d) from folding line F/S and the same distance (h) below the horizontal folding lines (T/F and T/S).

The simplest transfer occurs when a point is located in the top and side views and is transferred to the front view. This type of transfer is shown by the thin lines with the arrows in Figure 3-19b.

PROJECTION OF A LINE

There are different types of lines, such as normal, inclined, oblique, and curved. The same principle of point projection can be extended to line projection.

A **normal line** is parallel to two of the projection planes and perpendicular to the other one. Assume a horizontal line in the glass projection box. The line in the front view as shown in Figure 3-20 is parallel to the folding line and is the correct length. Since the line is parallel to the front projection plane, the line is true length (TL) in the front view. The line in the top view must be directly above the front view. The line in the top view is also true length because it is parallel to the top projection plane. Once the front and top views are sketched, the projection to the third view is fixed. Step I is to project a horizontal line from the front view to the side view. Step II is to project a horizontal line from the top view to the 45° line, and Step III is to project down to the side view. The intersection of these two projections is the point labeled 1,2. In this case the line is perpendicular to the side projection plane and, therefore, appears as a point in the side view. Obviously, it is not a true-length projection in the side view. When a line is parallel to two of the three projection planes, it will appear true length in two of the views and as a point in the third.

In the preceding paragraph the normal line was projected from the front and top views to the side view. If the normal line is located first in the front view and then in the side view, the line is projected to the top view as illustrated in Figure 3-21a. The normal line (appears as a point) in the side view is located a distance (d) from the folding line F/S and a distance (h) below the horizontal folding lines (T/F and T/S). Projection to the top view is illustrated by the thin lines with arrows.

Similarly, if a normal line is given in the top view, it is located in the side view a distance (h) from folding line T/S and a distance (d) from folding line F/S. Projection to the front view is shown in Figure 3-21b by the thin lines with arrows.

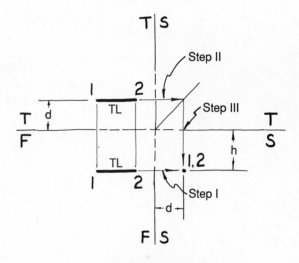

Figure 3-20
Projection of a normal line to the side view.

(a)

(b)

Figure 3-21
Projection of a normal line to the top and front views.

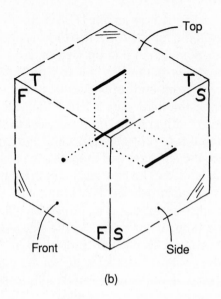

(a)   (b)

Figure 3-22
Normal lines perpendicular to the
top and front planes.

A normal line can be parallel to the front plane and side plane as shown in Figure 3-22a. Also, it can be parallel to the top plane and to the side plane as shown in Figure 3-22b. The projections of these lines are left as an exercise.

Now let's consider a line that is parallel to only one side of the glass projection box, say the front projection plane as shown in Figure 3-23. This line is called an **inclined line** and is parallel to only one

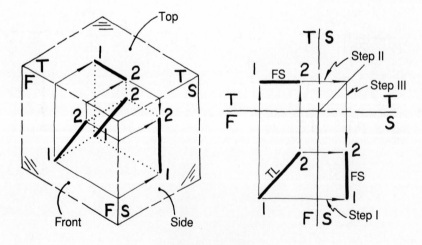

Figure 3-23
Projection of an inclined line
to the side view.

projection plane. The line is sketched in the front view with the correct slope and length. The ends of the line are projected to the top view. Since the line is parallel to the front projection plane, it appears as a foreshortened (FS) line in the top view. The line is projected to the side view by steps I to III in the same manner as in Figure 3-20; however, in this case the line does not appear as a point but as a foreshortened line.

In the preceding paragraph the inclined line was projected from the front and top views to the side view. If the inclined line is located first in the front view and then in the side view, the line is projected to the top view as illustrated in Figure 3-24a. The inclined line in the side view is located a distance (d) from the folding line F/S and a distance (h) below the horizontal folding lines (T/F and T/S). Projection to the top view is illustrated by the thin lines with arrows.

Similarly, if an inclined line is given in the top view, it is located in the side view a distance (h) from folding line T/S and a distance (d) from folding line F/S. Projection to the front view is shown in Figure 3-24b by the thin lines with arrows.

Lines that are parallel to only one of the three planes appear as true length in one of the views and foreshortened in the other two views. If the inclined

(a)

(b)

Figure 3-24
Projection of an inclined line to the
top and front views.

(a) (b)

Figure 3-25
Inclined lines parallel to the
top and front planes.

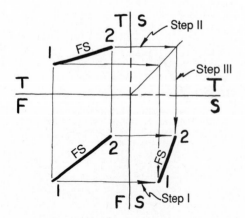

Figure 3-26
Projection of an oblique line
to the side view.

line had been parallel to the top projection plane as shown in Figure 3-25a, it would have appeared true length in the top view; if the line had been parallel to the side projection plane as shown in Figure 3-25b, it would have appeared true length in the side view. The three-view sketches of these two situations are left as exercises.

An **oblique line** is neither parallel nor perpendicular to any of the projection planes. The projection of an oblique line from the front and top views to the side view is shown in Figure 3-26. If the oblique line is located first in the front view, the ends of the line are projected to the top view. The location of ends 1 and 2 must be the correct distance from the front plane. Projection to the side view is shown by the thin lines with arrows.

Projection of an oblique line from the front and side views to the top view is shown in Figure 3-27a by the thin lines with arrows. Similarly, the projection of an oblique line from the top and side views to the front view is shown in Figure 3-27b.

It is important for you to know that when a line is not parallel to any of the projection planes, the line is foreshortened in the front, top, and side views. To obtain the true length of this type of line, an auxiliary view is necessary. Auxiliary views are discussed in a later chapter.

(a)   (b)

Figure 3-27
Projection of oblique lines to the
top and front views.

PROJECTION OF SURFACES

Although the study of projection of points and lines is important to understand the concept of orthographic projection, objects are actually constructed or manufactured by surfaces. When visualizing objects, the technically trained person thinks in terms of surfaces. An object can be made up of normal, inclined, oblique, and curved surfaces.

A **normal surface** is a plane surface of any shape that is parallel to a projection plane. The normal surface appears true size and shape in the plane to which it is parallel and a vertical or horizontal line on the other two projection planes. Many normal surfaces are rectangles; however, the shape could be triangular, circular, octagonal, and so forth. The projection of the normal rectangular surface, as illustrated in Figure 3-28, is identical to the projection of four normal lines. First, sketch the rectangle in the front view (since it is parallel to that view). Next, the rectangle is located in the top view directly above the front view. It will appear as a line and can be located a distance (d) from the folding line. From there on the procedure is merely to project the four points of the rectangle to the side view and join them by lines. This particular normal surface is parallel to the front projection plane. Normal surfaces

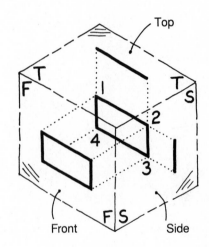

Figure 3-28
Normal surface projection.

(a)

(b)

Figure 3-29
Normal surfaces parallel to the
top and side planes.

are also parallel to the side or top projection planes. Figure 3-29a shows a normal surface (shaped as a triangle) that is parallel to the top projection plane. Figure 3-29b shows a normal surface (shaped as a hexagon) that is parallel to the side projection plane. Sketching these surfaces in the three views is left as an exercise.

An **inclined surface** is a plane surface of any shape that is perpendicular to one projection plane but inclined to the others. Figure 3-30 has the inclined surface perpendicular to the front projection plane. The procedure for the projection of the surfaces is similar to the procedure for a normal surface. It relies on the concepts given previously in the projection of a point and line sections of this chapter.

This inclined surface is perpendicular to the front projection plane; an inclined surface can also be perpendicular to the side or top projection planes. Figure 3-31a shows an inclined surface (shaped as a triangle) that is perpendicular to the top projection plane. Figure 3-31b shows an inclined surface (shaped as a hexagon) that is perpendicular to the side projection plane. Sketching these surfaces in the three views is left as an exercise.

At this point in our discussion we discontinue the use of the glass projection box; you should be familiar with its principle. An **oblique surface** is

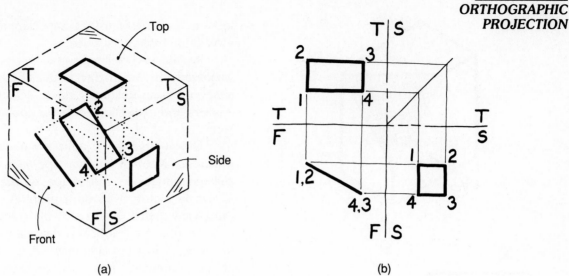

(a)

(b)

Figure 3-30
Inclined surface projection.

shown in Figure 3-32. It is a plane surface of any shape that is not perpendicular or parallel to any of the projection planes. The oblique surface does not appear true size in any of the views. To obtain the true size, auxiliary views are required. Fortunately, oblique surfaces do not occur too frequently in ob-

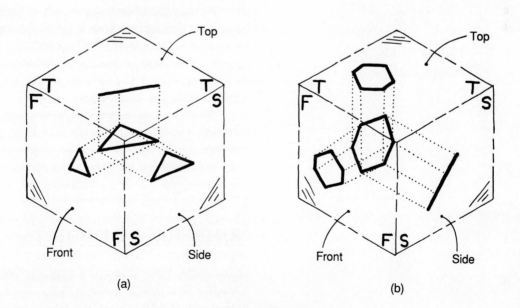

(a)

(b)

Figure 3-31
Inclined surfaces perpendicular to the
top and side planes.

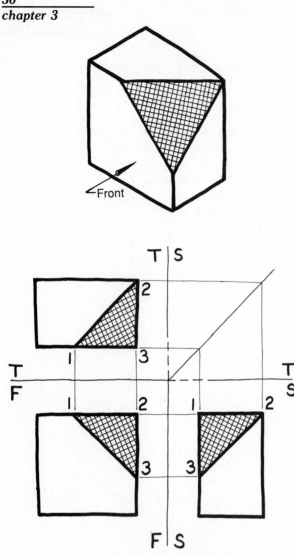

Figure 3-32
Oblique surface.

jects, and often it is not too important to have a true view of the oblique surface.

Normal and inclined surfaces can frequently be visualized and sketched without elaborate projection techniques. An oblique surface, however, requires analysis and projection of the endpoints of the surface.

Objects frequently have a number of **curved surfaces.** Material-removal operations and the resulting changes in the shape of the object are shown in Figure 3-33 by a pictorial sketch and a three-view sketch. Let's start with a rectangular solid and machine a cylindrical surface, as shown in Figure 3-33a. Note the side view where the line stops at the centerline. This type of projection occurs whenever one surface is tangent to another surface.

Now a small hole is drilled part way through the object, as shown in Figure 3-33b. Note that at the end of the drilled hole a triangle is formed by the end of the drill bit. Hidden or dashed lines are used for the drilled hole in the top and side view because these surfaces are not visible in that view.

The next step is to drill and counterbore two holes for cap screws, as shown in Figure 3-33c. Note that the bottom of the larger hole is a normal surface and shows a hidden line. Only one hole is shown in the side view because the other one is directly behind it.

Next we can machine another curved surface partway into the object, as shown in Figure 3-33d. Note how this surface shows up in the top and side view as a rectangle. If this surface had been machined to the end of the part, the long hidden line would have gone to the end of the part and the short hidden line eliminated.

When curves are irregular in shape, such as elliptical, exponential, involute, and so forth, it is necessary to project a number of points and then draw a smooth curve joining the points.

NUMBERING OF POINTS

Many times in complicated orthographic sketches we need to number points to determine their location in related views. There is no set standard to follow, but once the point is numbered, such as in the pictorial in Figure 3-34, there is only one proper

(a)   (b)

(c)   (d)

**Figure 3-33
Material-removal operations and
curved shapes.**

Figure 3-34
Numbering points.

place for it in the orthographic views. In the use of this technique, the point is numbered in two related views, and then located in the third view. Notice that the points will be in line vertically in the front and top views and in line horizontally in the front and side views. The best practice to follow is to use the same number in all views. Letters can also be used to identify points.

The numbers of all visible points are placed outside that view and invisible or hidden ones inside. For example, the point numbered 3 is visible in the front view, so it is located outside the view. In the top view that same point (number 3) is invisible and, therefore, located inside the view.

Let's take a more difficult object and see how the technique works. Suppose that we have the two views shown in Figure 3-35a. The numbers 1, 2, 3, and 4 have been used to identify the vertical surface on the right. The number 1 is at the top and front of the object; the number 2 is at the top and rear of the object; the number 3 is at the bottom and front of the object; and the number 4 is at the bottom and rear of the object. These four points are projected to the side view as shown by the arrows in Figure 3-35b and give the surface 1, 2, 3, 4.

We can now label another surface, say 5, 6, 7, and 8, the vertical surface on the left. Also we can label 9, 10, which forms a horizontal surface

(a)

(b)

Figure 3-35
Numbering points.

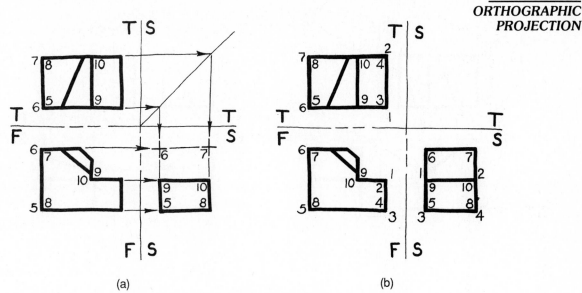

(a)   (b)

Figure 3-36
Numbering points.

with 1, 2. The projection of these points to the side view is shown in Figure 3-36. Where the lines intersect, the points 5, 6, 7, 8, 9, and 10 are located.

Wherever two points are joined in the front or top view, they must also be joined in the side view. The joined points are shown in Figure 3-36. Note that since points 6 and 10 are not joined in either the front or top views they are not joined in the side view.

Now we have some type of oblique surface that has not been labeled. Let's label the front view with the letters a, b, c, and d, as shown in Figure 3-37a. For the top view there is only one possible location for points a and b. For c and d there are two possible locations in the top view; however, since d is visible and lower than c in the front view, it must be closest to the projection plane in the top view. The points a, b, c, and d are projected to the side view as shown in Figure 3-37b. By joining the respective points together, the complete orthographic projection is obtained.

Figure 3-38 shows a pictorial of the object. Of course, if the pictorial had been given at the beginning, the numbering system would not have been necessary.

Curved surfaces are frequently located by projecting their centerlines. However, it is sometimes necessary to project points on the curved surface

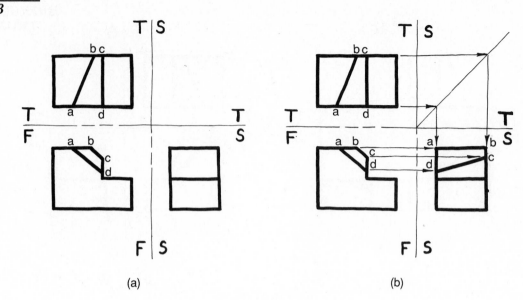

(a)   (b)

Figure 3-37
Numbering points.

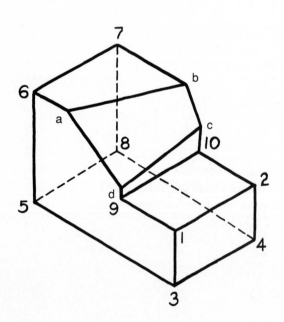

Figure 3-38
Numbering points.

and then join the points together. In Figure 3-39a we have a cylinder with an inclined surface that will project to the side view as an ellipse. The points 1, 2, 3, and 4 are labeled where the centerlines cross the circle and are projected to the side view by the thin lines with arrows as shown in Figure 3-39b. Additional points as shown by the letters a, b, c, and d may be necessary to obtain a smooth curve. For clarity the projection of these points is not shown in the figure. The rest of the view can be completed by visualization or projecting the points 5, 6, 7, and 8.

In our discussion we have concentrated on projection to the side view. If the front and side views are given, we could just as easily project the numbers to the top view. Or if the top and side views are given, we could project the numbers to the front view.

PRECEDENCE OF LINES

When two types of lines are coincidental or overlapped one has precedence over the other. A visible line takes precedence over a centerline or hidden line (see a and b in Figure 3-40). A hidden line takes precedence over a centerline (see c).

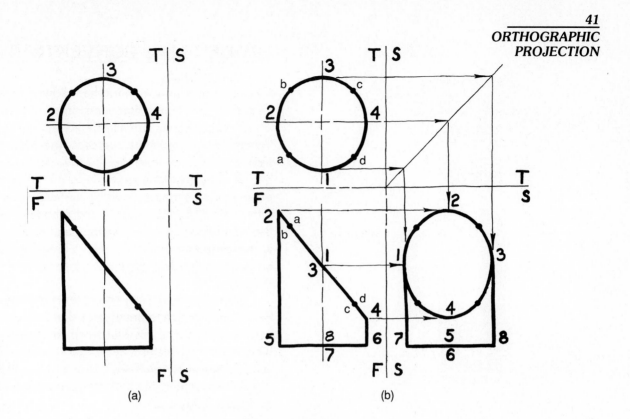

Figure 3-39
Projection of a curved surface.

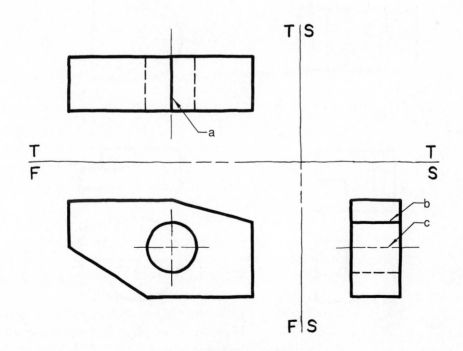

Figure 3-40
Precedence of lines.

HIDDEN-LINE CONVENTION

Certain conventions need to be followed when hidden lines join object lines and other hidden lines. These conventions are illustrated in Figure 3-41. Where hidden surfaces intersect, the hidden lines also intersect, as illustrated by the circles in Figure 3-41a. Where hidden surfaces do not intersect with other hidden surfaces or visible surfaces, there is no intersection, as illustrated by the triangles in Figure 3-41a. Hidden lines representing an arc should stop at the endpoints of the arc, with a break for the continuing straight-line part, as illustrated in Figure 3-41b.

Where a hidden line represents a continuation of a visible surface, a break should be left at the joint with the solid line, as illustrated in Figure 3-41c. This same principle as it applies to cylindrical objects is illustrated in Figure 3-41d. When two hidden lines are close together, they are staggered, as illustrated in Figure 3-41c.

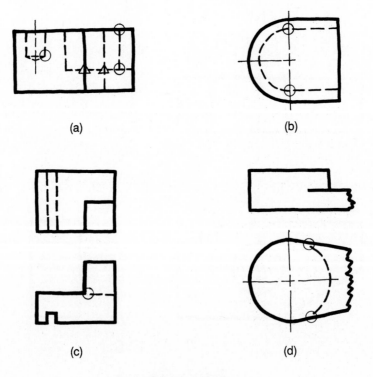

(a)

(b)

(c)

(d)

Figure 3-41
Hidden-line convention.

SUMMARY

In summary, one must know or do the following:

1. Orthographic projection means right-angle projection.

2. Orthographic projection uses projection planes.

3. Each set of two adjacent projection planes is perpendicular to each other.

4. The lines of sight (projectors) are parallel to each other and perpendicular to the projection plane.

5. The observer is assumed to be at an infinite distance from the projection plane.

6. A projection is the image of the object on the projection plane.

7. The three most commonly used projections are front view, top view, and side view.

8. The front view is the principal view.

9. Only one view of an object may be seen at one time.

10. Three views are usually necessary to describe an object; however, some objects might only need two or one.

11. Two of the planes of projection appear as an edge in each orthographic view.

12. Folding lines represent the edges or intersections of projection planes.

13. The object is stationary; it is never rotated.

14. A three-view sketch means that the observer has viewed the object from three different directions.

15. Choose the front view to show the most characteristic shape of the object.

16. The object should be in a stable position.

17. View the object to see as many visible lines in the front view as possible.

18. If only two views are needed, show the front and either the top or side.

19. The projection of a point is the foundation of orthographic projection.

20. A normal line is parallel to two projection planes and perpendicular to the other ~~two~~. *one*

21. An inclined line is parallel to only one projection plane.

22. An oblique line is neither parallel nor perpendicular to a projection plane.

23. A normal surface is parallel to one of the projection planes and perpendicular to the other two.

24. A normal surface is true size in one view and a vertical or horizontal line in the other two views.

25. An inclined surface is perpendicular to one projection plane but inclined to the others.

26. An inclined surface appears as a line in one view and a foreshortened shape in the other two.

27. Oblique surfaces are neither parallel nor perpendicular to any projection planes.

28. The numbering of points can be quite helpful in sketching views.

29. A visible line takes precedence over a hidden line and a centerline, and a hidden line takes precedence over a centerline.

30. Where hidden surfaces intersect, their hidden lines also intersect.

31. Where hidden surfaces do not intersect, their hidden lines do not intersect.

32. A break is left where a hidden line represents a continuation of a visible surface.

PICTORIAL
SKETCHING

chapter

4

Upon completion of this chapter, the student is expected to:

- Know the principles of isometric and oblique sketches.

- Be able to make an isometric sketch from a three-view sketch.

- Be able to make an oblique sketch from a three-view sketch.

- Be able to make an oblique and isometric sketch of a common object.

INTRODUCTION

An orthographic sketch is necessary to adequately represent and describe complicated and intricate parts and machines. This type of sketch is limited, however, to a person who has the special training necessary for its construction and interpretation.

Nearly everyone can read and interpret a photograph of an object. In those cases where intricate detail or interior construction is not needed, the pictorial form of sketch is adequate and very useful. Orthographic sketches show only two dimensions in a view. The various forms of pictorials show three dimensions in a view, thereby giving the object three-dimensional form. For this reason, a pictorial sketch has an important part in graphical representation. It becomes a means of rapid expression and explanation of ideas to the average person.

Our work will be confined to axonometric and oblique types of pictorial sketches, since they are the most convenient to make and the most commonly used.

PRINCIPLES OF AXONOMETRIC PROJECTION

Axonometric projection as illustrated in Figure 4-1 is similar to orthographic projection in that the observer is at an infinite distance, parallel lines of sight are perpendicular to the projection plane, and an image of the object is projected onto the projection plane. One of the differences between the two types of projections is the number of projection planes. Orthographic has three principal projection planes; however, axonometric has only one. Another principal difference is the position of the object. In orthographic projection, the object is in a stable position. In axonometric projection the object is in an unstable position, similar to a cube suspended on one point on a table.

There are three different types of axonometric projection. The type is determined by the position of the object. Since the object is rotated and tilted, a different relationship of the angle between the axes and the length along the axes is obtained.

Isometric projection occurs when the object is rotated and inclined to the projection plane in such a manner as to produce equal angles among the

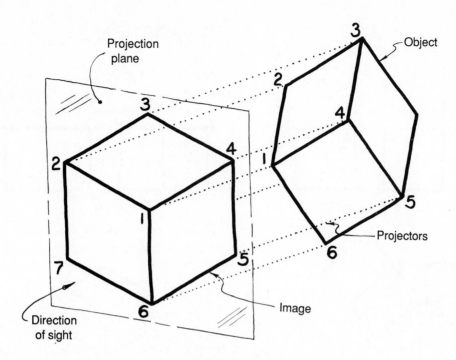

Figure 4-1
Axonometric projection.

three axes and equal lengths along each axis. Figure 4-2a illustrates an isometric projection of a cube. Dimetric projection (Figure 4-2b) occurs when the object is rotated and inclined to the projection plane in such a manner as to produce only two axes angles and two axes lengths that are equal. Trimetric projection (Figure 4-2c) occurs when the object is rotated and inclined to the projection plane in such a manner that none of the three axes angles or

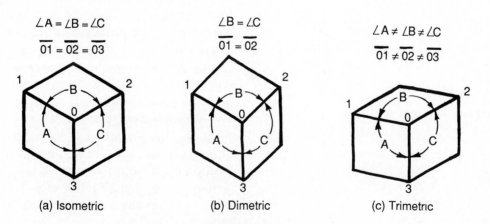

Figure 4-2
Types of axonometric projection.

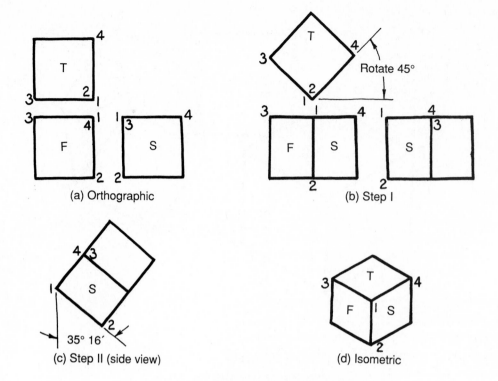

(a) Orthographic

(b) Step I

(c) Step II (side view)

(d) Isometric

Figure 4-3
Isometric projection by revolution.

lengths is equal. Because of the limited use of dimetric and trimetric in sketching, we will only study the isometric projection.

Isometric projection is the most widely used, since the same scale may be used on all axes. Figure 4-3 shows the development of an isometric projection. A three-view orthographic sketch of a cube is illustrated above. Assume that the cube is placed flat on, say, a level tabletop, with one surface parallel to the front projection plane that is perpendicular to the tabletop. To produce the isometric projection, rotate the cube to an angle of 45° with the front plane, viewed from the top (step I). The cube remains flat on the table. Next, tip the cube forward until the front corner 1, 2 makes an angle of 35° 16' with the front plane as viewed from the side (step II). This position gives the isometric projection of the cube as shown at (d) and establishes the isometric axes. One plane of projection, the front projection plane is used. The object is rotated and tipped to show the three surfaces of the cube in one view.

The resulting isometric axes are equal and at an angle of 120° to each other, as shown in Figure

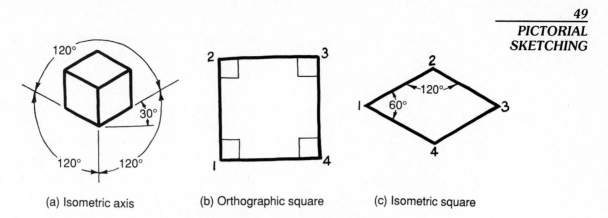

(a) Isometric axis   (b) Orthographic square   (c) Isometric square

**Figure 4-4
Square in isometric projection.**

4-4a. For simplicity, one of the axes is established in a vertical position which requires the other two axes to form an angle of 30° with the horizontal. Figure 4-4b shows a one-view sketch of a square surface. At each corner is a 90° angle, as shown at corners 1, 2, 3, and 4. In Figure 4-4c the square is shown in isometric form and the angles at corners 1 and 3 are at 60° and the angles at corners 2 and 4 are at 120°. In the orthographic sketch, the diagonals of the square are equal. In the isometric, the diagonals are unequal, neither being the true length of the diagonal. *Thus measurements can be made only along or parallel to an isometric axis.*

Since the object has been rotated and inclined at an angle with the projection plane, it will appear foreshortened when projected on that plane. A true isometric projection is foreshortened by 81.6%, as shown in Figure 4-5a. Special scales are available to construct an isometric projection. However, in practical work, an isometric sketch is constructed full size as shown in Figure 4-5b. Since the three isometric axes are at 120° to each other, the difference between an isometric projection and an isometric sketch is one of size, the isometric sketch being larger than a true projection.

ISOMETRIC SKETCHING TECHNIQUES

The most important sketching technique is the use of isometric paper. This paper must be turned so that the sloping lines make an angle of 30° with the

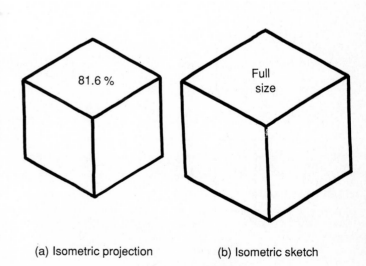

(a) Isometric projection   (b) Isometric sketch

**Figure 4-5
Isometric and ordinary scales.**

Figure 4-6
Isometric sketch using isometric paper.

horizontal axis, as shown by Figure 4-6. Lengths may be estimated by the grid on the paper. The paper is very helpful for sketching rectangular, sloping, or circular surfaces. Sketch lines should be dark and prominent. As noted in Chapter 1, special sketching pads of transparent paper are available that allow an isometric grid to be placed underneath the transparent paper. The isometric grid is visible through the transparent paper, and the resultant sketch has a more pleasing appearance without the isometric grid showing. Note the method of sketching the fillet. A round would be sketched similarly.

Figure 4-7 illustrates the isometric box construction technique. When working with objects with inclined surfaces, or cutouts, it is best to use the box construction method. To do this, merely show the outline of a box totally enclosing the object (step I); then make measurements along the edges of the box to locate the cutout corners or surface lines. Dimensions w, h, and d illustrate this technique (step II). The box should be developed using construction lines (3H-grade lead) and then sketched with prominent lines (HB-grade lead or a fiber pen) after the configuration has been established (step III).

Another technique is to build the piece up one block at a time, as shown in Figure 4-8. In isometric sketching, vertical measurements are made only on or parallel to the vertical axis. Horizontal measurements are made only on or parallel to the horizontal axes. In using the block method, first sketch the lower block. To locate corner 1 (step I), we must

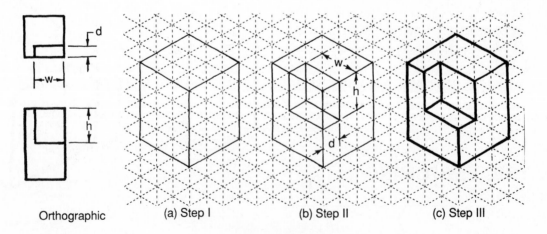

Orthographic   (a) Step I   (b) Step II   (c) Step III

Figure 4-7
Box construction method.

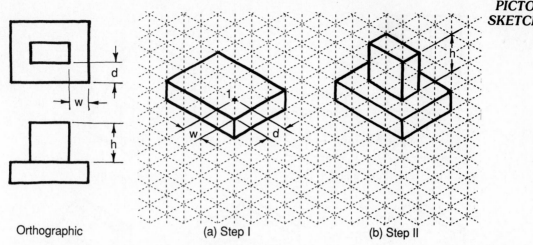

Orthographic　　　　(a) Step I　　　　(b) Step II

**Figure 4-8
Block method.**

make two horizontal measurements, w and d. We can now sketch height h and complete the upper block as shown in step II.

To sketch inclined surfaces, it is necessary to locate the end points of the surface, as illustrated in Figure 4-9 using the box-construction method. Step I locates the points 1 and 4 using one vertical measurement a and one horizontal measurement b, step II locates the points 2 and 3 using one horizontal measurement c, and step III completes the isometric by joining the points.

Orthographic　　　　(a) Step I　　　　(b) Step II　　　　(c) Step III

**Figure 4-9
Inclined surfaces.**

(a) Orthographic   (b) Steps I and II

(c) Steps III and IV   (d) Step V   (e) Isometric

Figure 4-10
Projection onto inclined surfaces.

Projection of features onto sloping surfaces requires the projection of sufficient points to outline the surface. In Figure 4-10 we have a square hole opening out onto an inclined surface. To show the hole opening on the inclined surface, construct a box enclosing the object in both the orthographic and isometric.

Step I.   Sketch in the projection of the opening on the end of the box 1234 in isometric.

Step II.   Project points 3 and 4 to the corner of the box, 5. Then project to point 6 on the sloping surface.

Step III.   Sketch a horizontal line on the sloping surface from 6.

Step IV.   Project points 3 and 4 horizontally back to the sloping surface to locate points 7 and 8.

Step V.   Use the same method to locate points 9 and 10.

Step VI.   Sketch in the complete outline of the object.

When using isometric paper, the grid lines make the construction much simpler.

(a) Orthographic   (b) Steps I, II, III, and IV   (c) Step V   (d) Step VI

**Figure 4-11
Isometric circle construction.**

Figure 4-11 illustrates the proper procedure to follow in constructing circles in isometric. It is known as the four-center ellipse method. Given is a square in orthographic with a circle on the surface of the square.

Step I. Lay out the square in isometric.

Step II. Lay out the two centerlines of the circle on this surface.

Step III. On the centerline, lay out the four points for the radius of the circle.

STEP IV. Parallel to the centerline, sketch lines through these points, forming an isometric square that would totally enclose the circle.

Step V. From corners 1 and 2 of the parallelogram, sketch lines to the midpoints of the opposite sides. These lines intersect at points 3 and 4, which are the center of the two end radii. Points 1 and 2 are the two centers of the other radii.

Step VI. Complete the circle by sketching all four arcs.

When using isometric paper, many of these construction lines are available, provided the circle center is located properly. With practice one should be able to adequately sketch a circle in isometric with just the use of the isometric grid. Many designers use an ellipse template.

The preceding description is applied to the front surface of a cube. For the top and side surfaces the construction is the same, but the appearance is different, as shown in Figure 4-12. It is important to

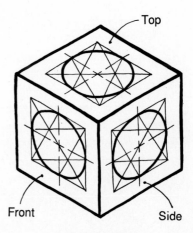

**Figure 4-12
Isometric circles on different surfaces.**

(a) Orthographic   (b) Isometric

Figure 4-13
Angles in isometric.

notice that the long diagonal of the parallelogram (the major axis of the ellipse) will slope upward to the right in the side surface, upward to the left in the front surface, and be horizontal in the top surface.

Angles in isometric are treated similarly to sloping surfaces. Figure 4-13 shows angles laid off in orthographic and the proper procedure for laying off angles in an isometric sketch. Distances along the isometric axes must be obtained in the orthographic view and measured along the isometric axes in the isometric sketch. The line, representing the angle, may then be drawn to this point. Angles may not be measured by their degrees in isometric sketches as these angles are bounded by nonisometric lines and are not true measurements.

Irregular shapes are treated as if they were a series of points with each point requiring three measurements in the isometric sketch. The technique is illustrated in Figure 4-14. The distances are obtained from the orthographic view, and measurements are made along or parallel to the isometric axes to establish a series of points. By joining these points, a smooth curve is constructed. Enough points should be located to ensure sketching a curve representing the contour shape of the object. The location of point 2 is illustrated by the distances w, h, and d. The plotting of points would also have been necessary for a round hole or an inclined surface rather than the square hole of Figure 4-10.

Because of the shape and location of some objects, a reversed isometric axis is sometimes used. Figure 4-15a shows the reverse position of the isometric axes in which we are looking up from the

(a) Orthographic   (b) Isometric

**Figure 4-14
Irregular shapes in isometric.**

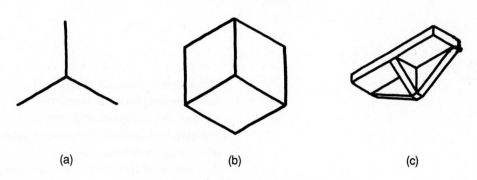

(a)   (b)   (c)

**Figure 4-15
Reversed isometric axes.**

bottom of the object rather than down from the top. In Figure 4-15b the basic cube is illustrated, and Figure 4-15c is a sketch of an object.

The orientation of the object should be accomplished to maximize the number of visible lines. Hidden lines are only used if the object is not adequately described by the visible lines. If necessary, the axes can be oriented so that none are vertical; however, they must be 120° to each other.

OBLIQUE SKETCHING

Oblique sketching is a form of pictorial representation that offers the advantage of easily illustrating objects having irregular shapes on one surface only. Oblique projection is not a true projection. While the observer is considered to be at an infinite dis-

(a) Orthographic   (b) Step I   (c) Step II   (d) Oblique

Figure 4-16
Oblique projection.

tance and the lines of sight are parallel, they are at an angle to the projection plane rather than perpendicular. All surfaces that are parallel to the projection plane remain in that configuration while the rest of the object is rotated and tilted.

Since oblique is not a true projection, the object is actually distorted. Figure 4-16a shows two orthographic views of an object. In Figure 4-16b (front and top views) the object is rotated; however, surfaces F, M, and R, which are parallel to the projection plane, remain parallel. This is analogous to taking a rectangle and converting it to a parallelogram. The object is tilted upward, keeping surfaces F, M, and R parallel to the projection plane as shown in Figure 4-16c, which is a side view. The completed oblique sketch is shown in Figure 4-16d.

For the general oblique sketch, the angle of the receding axis can be varied from 0° to 90° and the length to any scale we choose, as shown in Figure 4-17a. Since oblique sketching is not a true projection, these liberties can be taken. Two specific classifications of oblique sketches are constructed with a receding axis of 45° and are shown in Figure 4-17b and c. These are classified as cavalier and cabinet sketches. The scale on the receding axis of the cabinet is half size and the cavalier scale is full size. By varying the angle of the receding axis, we can accent different surfaces. By changing the scale along the receding axis, we can lengthen or shorten the appearance of the object.

Any length

Full size

Half size

0° to 90°

45°

45°

(a) General oblique

(b) Cavalier

(c) Cabinet

**Figure 4-17
Types of obliques.**

Figure 4-18 shows a comparison of a cylinder sketched in both isometric and oblique. Notice the distortion that we get in both projections. The isometric piece looks flattened rather than cylindrical, and the oblique seems to get larger toward the far end of the piece. Oblique sketching has its purpose, but it is not the answer to all types of pictorial sketching, any more than isometric is.

Since the diagonal of a square grid is at 45°, it is best to use a 45° angle for the receding axis. The cabinet style usually has a more pleasing appearance than the cavalier, as shown in Figure 4-19. Therefore, cabinet-style oblique sketches are recommended.

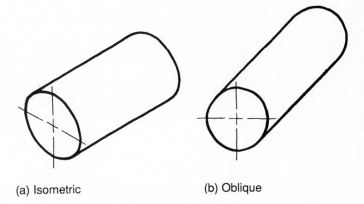

(a) Isometric

(b) Oblique

**Figure 4-18
Comparison of isometric and
oblique sketches.**

(a) Cavalier

(b) Cabinet

**Figure 4-19
Comparison of cavalier and
cabinet projections.**

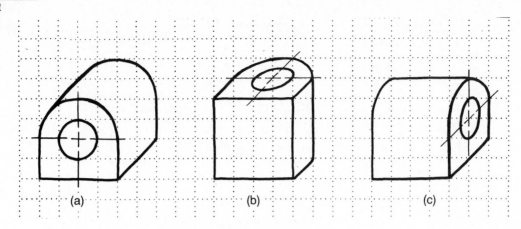

Figure 4-20
Selection of the front face.

The first and most important step in the construction of a cabinet-style oblique sketch is the selection of the front face. Circular or irregular surfaces of the object should be selected for the front face. Since the front surface of the object is parallel to the projection plane, it is a true orthographic front view and is constructed in the same manner. This fact is the basic reason for using an oblique sketch rather than an isometric sketch. Figure 4-20 illustrates the importance of the selection of the front face. The sketch shown in 4-20a is the correct one, since it is the easiest to sketch and has the most pleasing appearance. It is more difficult to construct the circles when the object is positioned as shown in Figure 4-20b and c.

Figure 4-21 shows the method of transferring centers along the receding axis. The circle is sketched in the front view; then, along the 45° axis

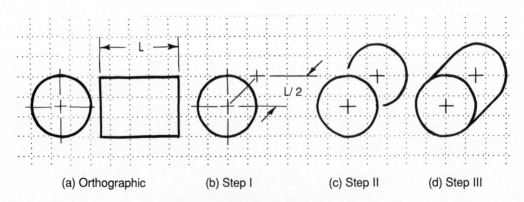

(a) Orthographic   (b) Step I   (c) Step II   (d) Step III

Figure 4-21
Transfer of centers along an axis.

(a) Orthographic

(b) Step I   (c) Step II

Figure 4-22
Steps for an oblique sketch.

of the center, a new center is established at the rear of the object equal to one-half the length of the object (step 1). The necessary portion of the rear circle is sketched in step II, and the sides of the cylinder are sketched tangent to the circles as shown in step III.

Steps for a more complicated object are shown in Figures 4-22 and 4-23. Figure 4-22 shows the

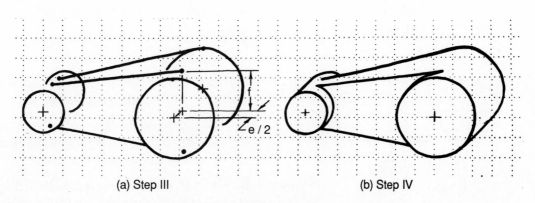

(a) Step III   (b) Step IV

Figure 4-23
Steps for an oblique sketch.

orthographic views and steps I and II. The first step is to sketch the large circle and the small circle. To locate the center of the small circle, dimension a is laid off on the 45° receding axis and then dimension b on the horizontal axis. Note that in the oblique sketch dimension a is one-half of dimension a in the orthographic view. Step II requires the location of the rear center of both cylinders and the sketching of these circles. Both dimensions c and d are laid off along the receding axis a distance equal to one-half of dimensions c and d in the orthographic view.

The location of the web of the casting is shown in Figure 4-23. Step III is the location of the eight tangent points of the web with the two cylinders. The location of one tangent point is shown by the dimensions e and f. After the other seven tangent points are located, the tangent lines are sketched. Actually, only five additional tangent points are necessary since one of the lines is hidden. Step IV requires the two circles at the web to be constructed and the entire sketch completed.

SUMMARY

In summary, one must know or do the following:

1. The three types of pictorial projection are axonometric, oblique, and perspective.

2. Isometric, diametric, and trimetric are the three types of axonometric.

3. An isometric projection is 81.6% of its full size.

4. In isometric sketching the object is rotated so that the angles between the three axes are equally spaced at 120°.

5. Measurements on all three axes use the same scale in isometric.

6. For isometric sketching, the object is turned to see as much detail as possible.

7. The distances (height, width, and depth) are transferred from the orthographic sketch along or parallel to the isometric axes.

8. Oblique is not a true projection because the line of sight is not perpendicular to the projection plane.

9. Oblique is beneficial in showing objects with irregular or circular shapes on one surface and on surfaces parallel to that surface.

10. Cavalier and cabinet oblique have 45° receding axes.

11. Cabinet oblique differs from cavalier in that the receding axis scale is one-half rather than full size.

12. Cabinet oblique has a more pleasing appearance than cavalier.

SECTIONAL VIEWS

chapter

5

Upon completion of this chapter, the student is expected to:

- Know the reason for using a sectional view.

- Be able to sketch a full-sectional view of an object.

- Be able to sketch a half-sectional view of an object.

- Know the sectioning rules and conventions.

- Identify the different types of sections, such as full, half, offset, revolved, and so forth.

INTRODUCTION

Sectional views are used to describe the interior construction and detail of parts and machines. The sectional view is particularly helpful in describing large or intricate pieces, such as an engine block, electric motor housing, and other objects where the exterior orthographic view becomes difficult to interpret because of so many hidden lines.

The concept of sectional views can be illustrated by slicing in half an apple. Figure 5-1 represents an apple that has been cut in half and one part removed. The interior features, including the seeds, core, and stem, can be clearly seen. This is not possible in a regular orthographic view without numerous hidden lines.

**Figure 5 - 1
Sliced apple.**

PRINCIPLE OF SECTIONAL VIEWS

The principle of taking sectional views is shown in Figure 5-2. The orthographic view of the object is shown at (a) and the cabinet oblique view of the object is shown at (b). An imaginary cutting plane is passed through the center of the object and one half of the object removed, as illustrated at (c). An arrow indicates the direction of sight, which is perpendicular to the cutting plane. The rear half of the object is behind the cutting plane. Where the cutting plane comes in contact with the object, the surface is cross-hatched. The orthographic front view and the sectional view as they would appear on the sketch are shown at (d). A cutting plane line is placed on the orthographic view to show the location of the cutting plane. At the end of the cutting plane line, arrows are drawn to indicate the direction of sight. Since more than one sectional view may be required to describe an object, letters are used for identification. In this case the letter A is used. The thickness of the cutting plane line is the same as a visible line. Construction of the cutting plane line is by a series of two short dashes and one long dash. The short dashes are approximately 3 mm (0.12 inch) long with a space between of approximately 1 mm (0.04 inch), and the long dashes can vary from 20 to 40 mm (0.8 to 1.6 inches). The cutting plane line is placed on the orthographic view that represents the cutting plane as an edge.

(a) Orthographic

(b) Oblique

Imaginary cutting plane

A ◄─── Cutting plane line

A ◄───

(c) Cutting plane

(d) Orthographic and sectional

SEC A-A

Figure 5-2
Sectional view theory.

The sectional view is shown next to the orthographic view. Note the crosshatched areas which show the exposed areas, and the label **SEC A-A** at the bottom of the view, which identifies that particular sectional view.

A sectional view is a special type of orthographic view that is better able to show the interior features of a complicated object. The imaginary cutting plane is identical to the imaginary reference plane of an orthographic view except that it is placed against the object after an imagined portion of that object is removed. Therefore, the only difference between the two types of views is in their technique of representation.

The section line technique is shown in Figure 5-3. Section lines are thin and are located approximately 3 mm (0.12 inch) to 5 mm (0.20 inch) apart. This spacing is not measured, but is estimated. Larger sketches permit greater spacing; smaller sketches permit closer spacing. The spacing should not be so close or so heavy as to make the crosshatched area look dark in comparison to the rest of the sketch. The outline or visible lines should stand out as the most prominent lines.

**Figure 5-3
Crosshatched.**

As a general rule, crosshatched lines are sketched at an angle of 45°; however, when the lines parallel the side of the object, the appearance is not too pleasing, as shown in Figure 5-3b. In such cases an angle of 30° as shown at (c) or 60° is more appropriate. The particular style of section lines illustrated in Figures 5-2 and 5-3 is the one for general use. It is also the style for cast iron. Section lining symbols for other materials are shown in Figure 5-4. It is usually unnecessary to sketch these different material symbols since the exact material

**Figure 5-4
Symbols for section lining.**

Incorrect   Correct

(a)   (b)   (c)

Figure 5-5
Visible and invisible lines
behind the cutting plane.

must be specified in precise terms. For this reason and to save time, the cast iron or general symbol is used on detail sketches. It may, however, be necessary to use the symbols for assembly sketches to show the different parts more effectively.

An important principle in the sketching of sectional views is to show all visible edges behind the cutting plane. The beginner will sometimes omit the visible lines behind the cutting plane as illustrated by the three missing visible lines in Figure 5-5b. We also have the poor practice of showing hidden lines in the sectional view. Only in special cases where hidden lines would save another view or lend clearness to the sketch are they shown.

Figure 5-5c shows the proper technique to follow in the construction of a sectional view. Notice that no hidden lines are shown and that all visible lines behind the cutting plane are shown.

CONVENTIONAL REPRESENTATION

Conventional representation or conventional practice is the term used to describe situations where the true projection is not sketched because it is difficult to accomplish and is not as easy to read. Figure 5-6 shows two sectional views of an object. The sectional view at (b) is a true projection, but is poor practice. The sectional view at (c) is the conventional presentation, and it is much easier to read and quicker to sketch. Conventional representation should be followed in making sectional views as well as the orthographic views described in Chapter 3.

The handling of solid parts and ribs or webs in section can result in a confused representation if a true projection is used. The general rule is that ribs, webs, spokes, and so forth are not crosshatched in sectional views. The omission of crosshatching the rib or web is to avoid the impression of a solid piece. A sectional view of a solid piece is shown in Figure 5-7a. In Figure 5-7b we have the ribs in section. A true projection of the ribs in the sectional view would result in the same view as the solid piece. Therefore, it is necessary to depart from a true projection and use the conventional representation shown.

Figure 5-8 shows the effect of webs in section. To properly convey to the reader the interior features, a conventional representation as shown at (b) is used rather than the true projection as shown at (c). Note the imagined position of the cutting plane line in the orthographic view as shown at (a) to aid in visualizing the correct method of constructing the sectional view. It is also important to note that the small holes are considered to be rotated.

The illustration in Figure 5-9a shows the correct handling of a solid-disk-type center of an object. From this sectional view, one would expect to find a solid center to the object, which it has. The illustra-

Figure 5-6
Conventional representation.

Figure 5-7
Ribs in section.

tion at (b) represents a spoke wheel, and the spokes in the section view are not crosshatched. If they had been crosshatched, the impression of a solid piece would be conveyed.

As a general rule, conventional practices are used to adequately convey the interior shape of the object.

Figure 5-8
Webs in section.

Figure 5 - 9
Flanges and spokes in section.

FULL AND HALF SECTIONS

Our discussion of sectioning has been concerned with the full-sectional view, which is the most common type. Another common type is the half-sectional view. Before discussing the half-sectional view, let's review the full section.

Figure 5-10 reviews the principle of taking a sectional view. An imaginary cutting plane is passed through the center of the object, and the part of the object in front of the plane is assumed to be removed as shown at (a). The cutting plane line,

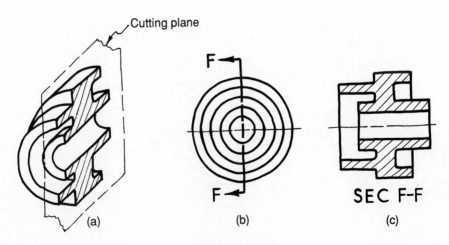

Figure 5 - 10
Full section.

which is the edge view of the cutting plane, is located in the front orthographic view. Arrows at the ends of the line indicate the direction of sight as shown at (b). Where the cutting plane comes in contact with the object, section lines are sketched in the sectional view. All visible lines behind the cutting plane are shown; however, hidden lines are not shown unless needed for clarity. The sectional view is identified by letter F, as is the cutting plane line.

Half-sectional views are used with symmetrical objects to show both the interior and exterior views as shown in Figure 5-11a. The cutting plane passes halfway through the object, and one-fourth of the object is assumed removed. Arrows indicate the direction of sight. They point in the direction the observer is looking, and therefore both must point in the same direction. This makes one of the arrows have a different appearance than the other one as shown at (b). Again, the sectional view must be identified and the cutting plane line shown. Hidden lines are normally not shown unless needed for dimensioning or clarity.

The object in Figure 5-11 is the same object as given in Figure 5-10; therefore, the differences between the two types of sectional views can be compared. Both the interior and exterior shapes of the object are clearly shown by the half-sectional view as illustrated in Figure 5-11c. Note that two of the vertical lines of the exterior portion and one of the vertical lines of the interior portion stop at the center line. This situation is quite common and makes visualization of half-section views difficult.

Cutting plane

(a)   (b)   SEC A-A   (c)

Figure 5-11
Half-section.

Cutting plane

A

A

Do not show
the offset in the
sectional view

SEC A-A

**Figure 5 - 12
Offset section.**

OTHER TYPES OF SECTIONAL VIEWS

There are other types of sectional views that you should be familiar with and be able to identify. These are offset, revolved, removed or detailed, broken and partial, and assembly. Each of these is applicable in particular cases and is discussed in the information that follows.

Offset

To convey the interior features of an object adequately, it may be necessary to change the direction of the cutting plane line. When this is done, we have an *offset section*, as illustrated in Figure 5-12. The section is identified by arrows and letters, and the line is offset according to the desired detail. The sectional view does not contain the change of direction of the cutting plane line. It is important in an offset section to make the cutting plane line dark and visible.

(a)   (b)   (c)

Figure 5-13
Revolved section.

Revolved

To save the construction of a full-sectional view or another orthographic view, a *revolved section* is used as shown in Figure 5-13. The sectional view is produced by a cutting plane that is revolved 90° about an axis at the center of the revolved section as shown at (a). Lines from the exterior view of the object should not cross the sectional view, as shown at (b). In some cases, such as the extrusion at (c), break lines are used. Two advantages of revolved sections are the elimination of the cutting plane line and of the sectional view designation. In addition, the sketch is frequently easier to understand and a view is saved.

Removed or Detailed

Removed or detailed sections are used to show the shape and construction of the object at a specified location. This type of sectional view is similar to revolved sections in that its purpose is to show the cross-sectional area. These removed sections are often sketched to enlarged scale to permit clear interpretation of the part construction. They must clearly show where the section was taken and the direction of sight. A number of removed sections in sequence can be used to show the change in shape of an object. An example of a removed or detailed section is shown in Figure 5-14. The sectional view can be located anywhere on the sketch in contrast to locating it as one of the principal views as shown in the figure. An excellent example of this type of sectional view would be a number of removed or detailed sectional views to describe the shape of an airplane wing or fuselage.

SEC A-A   SEC B-B

Figure 5-14
Removed or detailed section.

Broken and Partial

If only a small portion of an object is to be shown, a *broken and partial section* may be desirable. By this means, important interior construction of a small portion may be shown without full sectioning of the part or another view. An example of usage would be cross-drilled holes for bearings, locking setscrew holes, and threaded sections. Figure 5-15 illustrates an excellent example of use of this type of sectional view.

Assembly

Our last type of sectional view is the *assembly section,* which is shown in Figure 5-16. An assembly section is used to show the working relationship of parts in the unit and serves as a place to call out identification numbers and names. All areas of the same detail part should have the same section lining sloping at the same angle. Where the cutting plane passes along the center of bolts, nuts, pins, keys, spokes, ribs, or webs, these items should not be crosshatched in the sectional view. They should, however, be crosshatched if the cutting plane passes

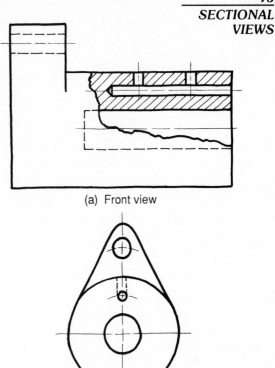

(a) Front view

(b) Side view

**Figure 5-15
Broken and partial sections.**

I. NUT
2. WASHER
3. WHEEL
4. BUSHING
5. PULLEY
6. SHAFT
7. BASE

**Figure 5-16
Assembly section.**

perpendicular to their axis. Section lines of adjacent parts should be crosshatched in opposite directions. The identification number permits reference to the detail drawing or catalog illustration for full information on the part. Note the use of different material symbols for the crosshatching.

SUMMARY

In summary, one must know or do the following:

1. Sectional views are produced by an imaginary cutting plane passing through the object.

2. The sectional view is the portion of the object behind the cutting plane.

3. The observer is at an infinite distance.

4. A sectional view is used to show the interior shape of an object.

5. Only material in contact with the cutting plane is crosshatched.

6. The arrows on a cutting line point in the direction in which the observer is looking.

7. The cutting plane line and sectional view must be identified.

8. Crosshatch lines should be spaced approximately 3 to 5 mm (0.12 to 0.20 inch) apart.

9. The crosshatched area should not appear dark on the sketch.

10. Hidden lines are not shown in the sectional view.

11. All visible lines behind the cutting plane are shown.

12. A sectional view is accompanied by all necessary exterior views.

13. A half-section is used only with a symmetrical object.

14. A half-section shows both the interior and exterior of the object.

15. Conventional representation, which is not a true projection, is used to make the sketch easier to read and much quicker to construct.

16. In an offset sectional view the cutting plane line changes direction.

17. A revolved sectional view is quicker to construct and easier to read than a full sectional view.

18. Removed or detailed sectional views are used to show the change in shape of an object such as an airplane wing.

19. Interior construction of a small portion is shown by a broken and partial sectional view.

20. An assembly sectional view shows the working relationship of the parts.

AUXILIARY VIEWS

Upon completion of this chapter, the student is expected to:

- Know the reasons for using auxiliary views.

- Understand the concept of primary auxiliary views.

- Be able to obtain the true length of an oblique line.

- Be able to obtain the true size of an inclined surface.

- Understand the concept of secondary auxiliary views.

- Be able to obtain the true size of an oblique surface.

chapter

6

INTRODUCTION

The principal views (front, top, and side) are sometimes inadequate to fully describe inclined or oblique surfaces. For manufacturing purposes these surfaces must be shown in their true size for dimensioning and visualization. For this reason we need to expand our knowledge beyond orthographic projection. As an example, the inclined surface of Figure 6-1 is not adequately described in the top view because the hole is an ellipse and the surface is not true size. Although the top view is a true projection, it is confusing.

The primary reason for using auxiliary views is to describe the shape of an object whenever it is not adequately described by the three principal orthographic views. Specifically, auxiliary views are used to show the true length of oblique lines; the true distance between points, lines, or planes; the true size, shape, and slope of surfaces; and true angles between lines or planes. Our study will be limited to the true length of oblique lines, the true size of inclined surfaces, and the true size of oblique surfaces. The other uses are beyond the scope of this book and are better left to a descriptive geometry textbook. Such constructions are not well sutied for sketching.

PRINCIPLE OF PRIMARY AUXILIARY VIEWS

A view obtained by projection on any plane except the six principal planes is known as an **auxiliary view**. A **primary auxiliary view** is projected onto a plane that is perpendicular to one of the principal planes and inclined to the other two principal planes. The observer is at an infinite distance, and the parallel lines of sight are perpendicular to the auxiliary plane. Based on this description, an infinite number of primary auxiliary views can be obtained. The location of the auxiliary plane is based on the purpose of the auxiliary view.

Figures 6-2, 6-3, and 6-4 show the auxiliary plane as it relates to the principal planes. In Figure 6-2 pictorial and orthographic views of the front, top, and auxiliary planes are shown. Both the auxiliary plane and the top plane are perpendicular to the front plane; therefore, we have a folding line

(a) Orthographic

(b) Isometric

**Figure 6-1
Inadequacy of principal views.**

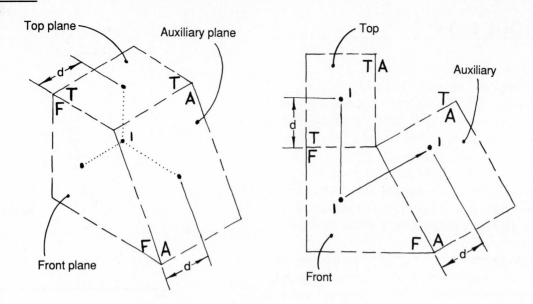

Figure 6-2
Auxiliary plane from the front view.

T/F between the front view and the top view and another one A/F between the front view and the auxiliary view. This is the basis for the construction of the auxiliary view. When an auxiliary view is taken off the front view, distance (d) from folding line T/F to point 1 in the top view is the same as distance (d) from folding line A/F to point 1 in the auxiliary view. Another way of saying the same

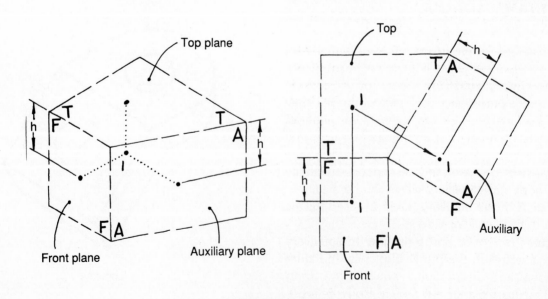

Figure 6-3
Auxiliary plane from the top view.

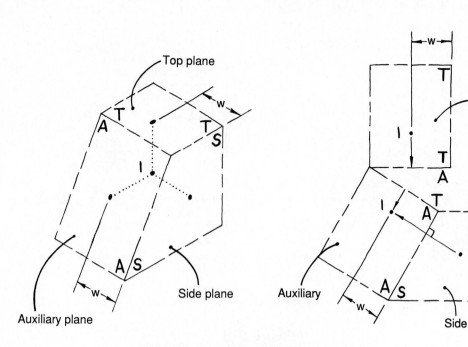

Figure 6-4
Auxiliary plane from the side view.

thing is, "when looking at the top view, folding line T/F represents the edge view of the front projection plane, and when looking at the auxiliary view, the folding line A/F also represents the edge view of the front projection plane; therefore, distances from these folding lines in their respective views must be equal."

Figure 6-3 shows the situation where the auxiliary view is taken off the top view. Since both the front projection plane and the auxiliary projection plane are perpendicular to the top projection plane, distances from folding lines T/F and A/T in the front and auxiliary views must be equal. This fact is illustrated by the distance h to point 1 in the front and auxiliary views. Figure 6-4 shows the situation where the auxiliary view is taken off the side view. Since both the top projection plane and the auxiliary projection plane are perpendicular to the side projection plane, distances from folding lines T/S and A/S in the top and auxiliary views must be equal. This fact is illustrated by the distance (w) to point 1 in the top and auxiliary views. Note that folding line A/T occurs twice in a manner analogus to folding line T/S.

To further illustrate the concept, an object is placed inside the projection box, as shown in Figure 6-5. The purpose of the auxiliary view is to show

Figure 6-5
Auxiliary view from the front view.

the true size of the inclined surface. To do this, an auxiliary plane is located parallel to the inclined surface and perpendicular to the front projection plane. The distance *d* is the distance of the object back of the front plane as shown in the pictorial, the auxiliary view, and the top view. All points, lines, and surfaces must be the same distance from folding lines T/F and A/F in the top and auxiliary views, respectively. The projectors, as illustrated by the thin lines with arrows, are perpendicular to folding line A/F. The association of the orthographic and the pictorial at this time is very helpful.

Figure 6-6 shows another inclined surface; however, the auxiliary view is taken off the top view. The distance h is the same in the front view as in the auxiliary view because both are measured as a vertical distance below the top plane. All points, lines, and surfaces in the front and auxiliary views must be the same distance from the top projection plane. The projectors, as illustrated by the thin lines with arrows, are perpendicular to folding lines A/T. Remember that folding lines T/F and A/T are the edge views of the top projection plane in their respective views.

Similarly, Figure 6-7 shows an inclined surface that requires the auxiliary view to be taken off the side view. The distance w is the same in the top view as in the auxiliary view because both are measured as a horizontal distance to the left of the side

Figure 6-6
Auxiliary view from the top view.

plane. All points, lines, and surfaces in the top and auxiliary views must be the same distance from the side projection plane. The projectors, as illustrated by the thin lines with arrows, are perpendicular to folding line A/S. Remember that folding lines T/S and A/S are edge views of the side projection plane in their respective views.

Figure 6-7
Auxiliary view from the side view.

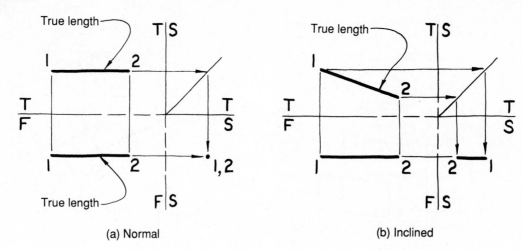

(a) Normal   (b) Inclined

Figure 6-8
True length of normal and inclined lines.

TRUE LENGTH OF A LINE

Many technical problems require the finding of the true length of a line. The true length of normal and inclined lines is given in one of the principal views, as illustrated by Figure 6-8. The information was given in Chapter 3 and is presented at this time as

(a) Steps I and II   (b) Step III

Figure 6-9
True length of an oblique line
from the top view.

a review. However, the true length of an oblique line must be found by an auxiliary view, as shown in Figure 6-9. In step I an auxiliary plane is constructed parallel to one view of the line. It does not make any difference whether the top, front, or side view is used. In this case the top view is used. The edge view of the top plane as seen in the auxiliary view is designated by the folding line labeled A/T. In step II the points 1 and 2 are projected perpendicular to that folding line. Step III consists of measuring the respective distances from the points 1 and 2 to the folding line T/F. These distances, a and b, are transferred to the auxiliary view to locate the points 1 and 2. The line joining these points is true length. Remember that the folding lines A/T and A/F represent the edge view of the top plane in the auxiliary and front views, respectively.

Note that the side view was not used; however, it could have been because the distances a and b are given in the side view, too. Figure 6-10 shows the construction of the true length of a line when the auxiliary is taken off the front view. Note that the distances a and b can be obtained from either the top or side view. Figure 6-11 shows the construction of the true length of a line when the auxiliary is taken off the side view. Note that the distances a and b can be obtained from either the top or front view.

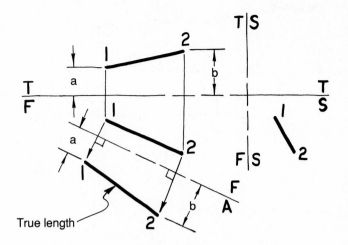

Figure 6-10
True length of an oblique line
from the front view.

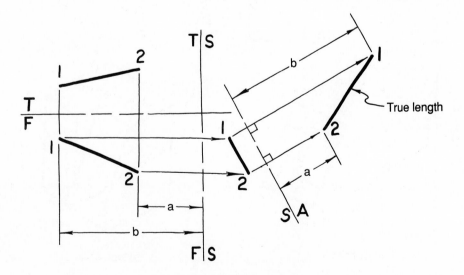

Figure 6-11
True length of an oblique line
from the side view.

TRUE SIZE OF AN INCLINED SURFACE

Another important use of auxiliary views is to obtain the true size of an inclined surface. Figure 6-12 shows the method of obtaining the true size of an inclined surface. The method is similar to that used to obtain the true length of a line. In step I the auxiliary plane, as represented by folding line A/F, is established parallel to the edge view of the inclined surface, which is labeled by the numerals 1, 2, 3, and 4. In step II projectors of the points 1-2 and 3-4 are constructed perpendicular to the folding line A/F. The distances a and b are used to locate the points on the projectors (step III). Lines are constructed between the points as shown at (d) and the remaining points of the object are located in the same manner, thereby giving the completed auxiliary view. The true size of the inclined surface is shown in the auxiliary view. An isometric projection is given at (e) for reference purposes only.

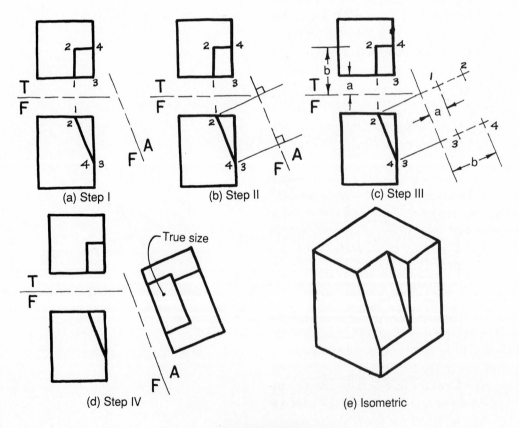

Figure 6-12
True size of an inclined surface
from the front view.

Figure 6-13
True size of an inclined surface
from the top view.

Figure 6-13 shows the construction of the true size of an inclined surface taken off the top view. Similarly, Figure 6-14 shows the construction of the true size of an inclined surface taken off the side view.

It is not always necessary to show the complete auxiliary view. In Figure 6-15 only a partial auxiliary view is needed because the center of the block is adequately described in the top and front views. The inclined surfaces are shown in detail in the two auxiliary views. In most cases we do not attempt to add the true projection in the top view because this would be time spent needlessly.

Figure 6-16 illustrates the true size of inclined surfaces for dimensioning and visualization. An object of this type would not require all the detail in the top view of the holes on the inclined surfaces. Dimensioning and location of the holes on the inclined surfaces would be done in the auxiliary views. Unless needed for clarity, it is not necessary to construct the complete view of all the inclined surfaces. In fact, a complete auxiliary view can be confusing. Partial auxiliary views are common practice.

Figure 6-14
True size of an inclined surface
from the side view.

Figure 6-15
Partial auxiliary views of
inclined surfaces.

Figure 6-16
Partial auxiliary views of
inclined surfaces.

TRUE SIZE OF AN OBLIQUE
SURFACE

The problem of finding the true size of an oblique surface, the surface 1-2-3 in Figure 6-17, is somewhat more difficult. An oblique surface does not show as an edge in any of the three principal views. Therefore, we will need to take two auxiliary views to show an oblique surface in true size. This additional auxiliary view, called a **secondary auxiliary view,** is projected from a primary auxiliary view on a projection plane that is inclined to all the principal projection planes. An infinite number of secondary auxiliary planes is possible; however, only one will achieve the desired result.

Figure 6-17 shows the method of obtaining the true size of an oblique surface. The steps are as follows:

Step I. Obtain the true length of a line on the oblique surface in one of the principal views. A horizontal line 1-4 is constructed on the oblique surface 1-2-3 of the front view. Since the line is parallel to the top projection plane (folding line T/F), the view of the line in the top view 1-4 is true length. Note that in some cases such as Exercise 6-4 the true length of the line is part of the surface and does not need to be constructed.

Step II. An auxiliary plane as presented by folding line A/T is established perpendicular to the true length line 1-4 in the top view.

Step III. The auxiliary view of the oblique surface is a straight line; therefore, we are seeing the edge view of the surface 1-2-3. The method of obtaining this view is the same as previously described under inclined surfaces. Distances of a, b, and c from folding line A/T are equal to the distance of a, b, and c from folding line T/F and represent the distances from the top projection plane.

Step IV. The secondary auxiliary plane is established parallel to the edge view of the oblique surface and is represented by folding line A/SE.

Step V. Projectors of points 1, 2, and 3 are constructed perpendicular to folding line A/SE. Distances from folding line A/SE to points 1, 2, and 3 are equal distances from folding line A/T to points 1, 2, 3 and are represented by the letters e, f, and g.

(a) Step I

(b) Steps II and III

(c) Step IV

(e) Isometric

Figure 6-17
True size of an oblique surface
from the top view.

Step VI. The points 1, 2, and 3 are joined to give the true size of the oblique surface. Usually, the rest of the object is not completed because it is adequately represented in the other views.

In this case the auxiliary views were projected from the top view. The same results could have been obtained by projecting from the front or side views and these cases are left as exercises.

Sketching auxiliary views requires great care. The reader may find it desirable to use a small triangle and scale.

SUMMARY

In summary, one must know or do the following:

1. Auxiliary views describe the shape of an object when the principal views are inadequate.

2. An auxiliary plane is located in the desired position and the observer is at an infinite distance.

3. Projectors are perpendicular to the auxiliary plane, and one of the principal planes is seen as an edge.

4. Distances are measured in a previous view showing the edge of the same plane as the auxiliary view.

5. The true length of an oblique line is obtained by an auxiliary plane parallel to any view of the oblique line.

6. The true size of an inclined surface is obtained by an auxiliary plane parallel to the edge view of that surface.

7. Partial auxiliary views are a common practice.

8. The true size of an oblique surface requires a primary auxiliary view and a secondary auxiliary view.

9. The first auxiliary for an oblique surface produces an edge view of that surface.

DIMENSIONING

chapter

7

Upon completion of this chapter, the student is expected to:

- Know the conventions of good dimensioning practice.

- Be able to dimension a three-view sketch.

INTRODUCTION

Our previous discussions in this textbook have centered on the shape or contour of the object as described by the various views (shape description) of an object. In this chapter and Chapter 8, we will be concerned with the size of the object and with the location and form of geometric features of the object. Dimensions are used to supply operating personnel with the necessary information (size, form, and location) to manufacture the object. They are not necessarily the same dimensions used to create the shape description. The object must be fully dimensioned so that operating personnel are not required to add, subtract, or multiply values or to use a scale to determine a measurement.

When applying dimensions, one should remember the following important criteria:

1. Accuracy: the values are correct.

2. Clearness: most appropriate position.

3. Completeness: no omissions.

4. Readability: lines, values, and notes are legible.

The information in this chapter is based on dimensioning practices as defined by ANSI Y14.5M 1982.

NOMENCLATURE

A **dimension** line is a thin line with a break near its center for insertion of the dimension value. It is terminated at both ends with an arrowhead whose length, 3 mm (0.12 inch), is approximately three times its width, 1 mm (0.04 inch). The arrowhead touches an extension line, which indicates the relationship of the dimension line to the object. Extension lines begin approximately 1.5 mm (0.06 inch) from the object line and extend approximately 3 mm (0.12 inch) beyond the arrowhead. Figure 7-1a illustrates the correct method of constructing these lines. The lines are thin, and are usually sketched with a 3H-grade pencil. The correct method of constructing the arrowhead is shown at (b).

Figure 7-2 illustrates the space between dimensions. As a general rule, the first dimension line is placed 10 mm (0.40 inch) from the object and all

Figure 7-1
Dimension and extension lines.

Figure 7-2
Space between dimensions.

succeeding dimensions are placed 6 mm (0.24 inch) apart.

Because of congested spaces, special techniques are necessary to enable the dimensions to be readable. Figure 7-3a illustrates the special technique for extension lines, and at (b) the techniques for dimensioning small values are shown.

Except in special cases, object lines or hidden lines should not be used as extension lines; however, as shown in Figure 7-4a, centerlines are frequently used as extension lines. Extension lines can cross other extension lines, object lines, hidden lines, or centerlines. In these cases, no gaps occur in the extension line at the crossing point. It is not considered good practice for an extension line to cross a dimension line. Proper ordering of the dimensions from small to large can avoid that situation.

Remember that the object and hidden lines must dominate so that the object stands out. It is not considered good practice to have dimension lines and values within the views of the object as shown at Figure 7-4a. Dimensioning to hidden lines is to be avoided; however, this is not always possible (see right-side view of Figure 7-6). These conventions cannot be followed at all times; therefore, judgment is usually required.

Another dimensioning technique is the use of **leaders**. They are used to direct a dimension, symbol, or local note to its intended place on the draw-

Figure 7-3
Dimensioning in crowded spaces.

Figure 7-4
Crossing other lines.

ing as illustrated in Figure 7-5. Local notes apply to a specific operation and are placed as near as possible to the detail concerned. A straight line is sketched from the local note to the part feature with an arrowhead at the end. For circles, the straight line if extended would pass through the center of the hole. Local notes are printed horizontally. General notes differ from local notes in that they apply

(a)   (b)

Figure 7-5
Leaders and finish marks.

to the entire object. A typical general note would be ALL DIMENSIONS ±1 mm.

Finish marks are placed on all views of an object where the surface appears as an edge, as shown in Figure 7-5. The finish mark is V-shaped with an included angle of 60° and with the right side twice as high (approximately 7 to 8 mm) as the left side. It points toward the bottom of the page. Details such as the roughness height, waviness height, and lay of the surface quality may be incorporated into the symbol. These details are discussed in Chapter 8. Finish marks are omitted on features where a note specifies a machine operation. They should also be omitted on holes and surfaces of finished stock. Dimensioning should be to finished surfaces.

All dimension values and notes are read from the bottom of the drawing. This method of placing the dimensions is called **unidirectional**. The aligned method whereby the values and notes are read from the bottom and the right side is no longer recommended.

DIMENSIONING VALUES

At one time workers scaled the various views of an object to obtain dimensions. As the manufacture of products became more complicated, the previously described system of dimensioning was developed.

Initially, all dimension values were in whole numbers and fractions of an inch. However, as more precise measurements were needed, the use of fractions was replaced by the decimal inch and the millimeter systems. Fractions will continue to be used for certain commercial commodities that use standardized nominal designations, such as pipe, lumber, and fastener sizes.

When using millimeter dimensioning, the following practices are observed:

1. Where the value is less than 1 millimeter, a zero precedes the decimal point (0.3).

2. Where the value is a whole number, neither the decimal point nor a zero is shown (3).

3. The last digit to the right of the decimal point is not followed by zeros (3.3, not 3.300) unless the value is part of a tolerance specification.

4. Neither commas nor spaces are used to separate groups of digits such as thousands.

When using decimal inch dimensioning, the following practices are observed:

1. Where the value is less than 1 inch, a zero is not used before the decimal point (.25).

2. Zeros are added to the right of the decimal point to indicate accuracy (3.300, not 3.3).

3. Neither commas nor spaces are used to separate groups of digits such as thousands.

Dual dimensioning whereby both millimeters and inches are given is sometimes used; however, this practice is discouraged. There are three ways to convey this information: by position at the dimension line, by the use of brackets, or by a conversion chart.

Unless otherwise noted, all dimensions in this book are in millimeters.

DIMENSIONING FOR SIZE

To adequately dimension an object for size, the width, height, and depth of all geometric features should be specified. As a general convention (see Figure 7-6), dimensions for the height of the object

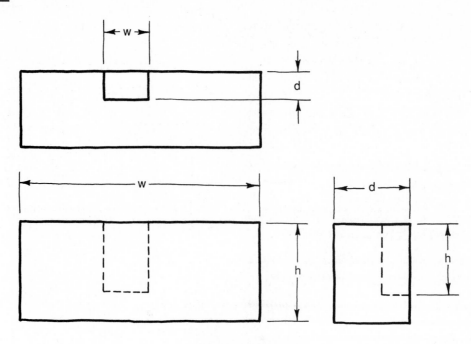

Figure 7-6
Placing dimensions.

should be placed between the views; the width should be placed above the front view; and the depth above the side view or to the right of the top view. As the object becomes more complicated than the one in the figure, this convention cannot be followed in all situations.

Another convention is to place dimensions adjacent to the view that shows the profile or contour of the geometric feature. This is illustrated by the dimensioning of the slot in the top view of Figure 7-6. You will note that the dimension for the height of the slot utilizes a hidden line; this is unavoidable. The location of the slot is not given—dimensioning for location is discussed in the next section.

Whenever possible, related dimensions should be grouped. Figure 7-7a illustrates one method of grouping dimensions. In this method there is a functional relationship of surface B and C to A. Figure 7-7b illustrates one method of grouping dimensions. In this method there is a functional relationship of surface B to A and surface C to B. The function of the object plays a critical role in determining the method to use.

Figure 7-7a also indicates a common error. When dimensions are in series, there is no need for the last dimension since the overall width of the object is given. The convention is to omit the least im-

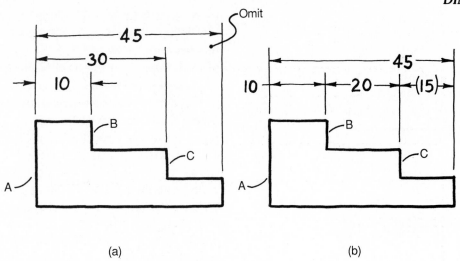

(a) (b)

Figure 7-7
Grouping dimensions.

portant dimension in the series. It could, however, be included as a reference, in which case the value is enclosed in parentheses.

Miscellaneous shapes have different methods of dimensioning, as illustrated in Figure 7-8. The most common shape is the cylinder, and it is dimensioned in the rectangular view as shown in Figure 7-8a. At (b) is a sphere with radius 9 mm. A cone and its method of dimensioning are shown at (c). A chamfer is dimensioned by two linear dimensions or

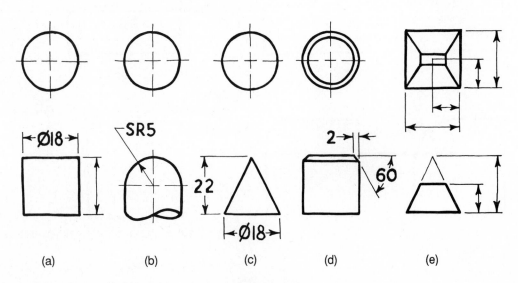

(a) (b) (c) (d) (e)

Figure 7-8
Shapes.

(a)

(b)

Figure 7-9
Angles.

by a linear and an angular dimension as shown at (d). When the angle is 45°, the chamfer may be dimensioned as shown in Figure 7-5b. The method of dimensioning a pyramid is shown at (e).

Angles of inclined surfaces utilize two methods of dimensioning, depending on the accuracy desired. The least accurate is shown in Figure 7-9a. At (a) the angle is specified by a linear value and an angular one expressed in either decimal degrees as shown or by degrees, minutes, and seconds. The coordinate method as shown at (b) is more accurate because the angle variation (when given in degrees) increases with the distance from the beginning point. Note that the dimension line for the angle is curved.

The dimensioning of curved surfaces has some specific conventions. Figure 7-10 illustrates the method of dimensioning fillets (top) and rounds (bottom). These shapes occur in various processes such as casting, welding, and milling. The letter R precedes the dimensioned value. If there are a number of fillets and rounds with the same radii, a local note such as 3× or a general note may be more appropriate than dimensioning each one. A typical note is ALL ROUNDS AND FILLETS R4.

Figure 7-10 can also be used to illustrate the general rules for fillets and rounds as well as some

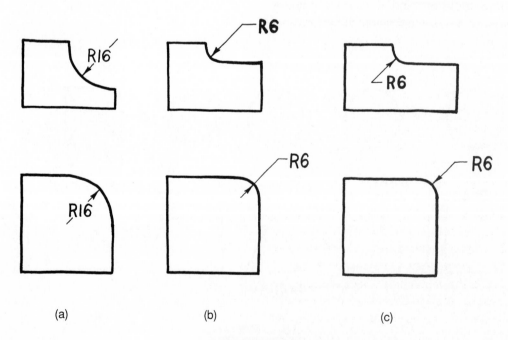

(a) (b) (c)

Figure 7-10
Dimensioning of radii.

Figure 7-11
Dimensionally located radius center.

of the rules for all radii. Where location of the center is important and space permits, a dimension line is sketched from the radius center with the arrowhead touching the arc, and the dimension is placed between the arrowhead and the center as shown at (a). If space is limited, the dimension line is extended through the radius center as shown at (b). Where it is inconvenient to place the arrowhead between the radius center and the arc, it may be placed outside the arc with a leader, as shown at (c). If the center of the radius is not dimensionally located, it is not indicated. In Figure 7-10 none of the centers is dimensionally located—they are controlled by other features, such as tangent surfaces.

Where a dimension is given to the center of a radius, by means of extension and dimension lines, a small cross is sketched at the center as illustrated by Figure 7-11a. If the dimension line is too long for the space available, it is foreshortened as shown at (c). The method of dimensioning partially rounded ends is shown at (b) and fully rounded ends at (d).

Figure 7-12 illustrates another method of dimensioning round end parts. The holes should be

Figure 7-12
Round end parts.

located between centers, and the radius of the end of the part given. Because operating personnel work with centers to locate arcs and holes, the overall dimensions are not normally given. Cylinders are dimensioned in their rectangular view, but holes are usually dimensioned by leaders. The depth of the hole is usually given in the local note using the symbol $\downarrow$ or printing DEEP *after* the value rather than in another view. This enables operating personnel to have the size and depth at the same place. The overall width of the object is 78 mm, and the height is 32 mm. If these values are desired, they are included as references. There are other methods of dimensioning holes and the reader is referred to the standard.

A very common geometric feature of an object is a hole for a bolt or screw. Figure 7-13 illustrates the methods of dimensioning these typical fastener holes. The callout note for a counterbored hole is illustrated at (a). First is the diameter symbol, which is followed by the hole size; next is the counterbore symbol (⊔), which is followed by its size; and last is the depth symbol and its value. Counterboring a hole is for the purpose of placing the head of the screw below the surface of the part. It is used largely for socket and fillister head screws.

The callout note for a countersunk hole is illustrated in Figure 7-13b. After the diameter symbol and the hole size, the letters THRU are given. The THRU symbol stands for "through" and is used wherever it is not obvious that the hole goes through. Next is the countersunk symbol (∨), which is followed by its diameter and included angle. The countersunk symbol is a wide V with an included angle of 90°. An × between the diameter and the angle is read as "by" and has a space before and after it. Therefore, × can be read as "times" (given previously) or "by." It is important to remember that the diameter is given and not the depth of the countersunk hole, as this is a very difficult and inaccurate measurement to make. Countersunk holes are used with flathead screws to place the head of the screw even with the surface of the part. Because the bearing surface of the head is at an angle, the countersunk hole may be used on thin parts, such as door hangers and thin covers.

The method of specifying a spotfacing operation is shown in Figure 7-13c. In this case there are three identical holes, as given by the 3× in the callout. The × is read as "times" and there is no space between the 3 and the ×. After the diameter of the hole, the spotface symbol (⨆) and its diameter are given. The last part of the symbol is the depth of the spotface. Note that the spotface symbol and the counterbore symbol are the same. The two are quite similar except that spotfacing has a larger diameter and its depth is only a few millimeters. If the depth is not given, the spotface operation is to the minimum

(a)   (b)   (c)   (d)

Figure 7-13
Typical holes for fasteners.

necessary to clean up the surface to the specified diameter; or, the depth of the remaining material may be specified in another view. Spotfacing is used to remove the burr left by the drilling operation and/or to provide a bearing surface for a washer or bolt head.

Counterdrilled holes, as shown in Figure 7-13d, are similar to counterbored holes except that the bottom surface has the same shape as a drill bit, which is usually at an included angle of 120°. It is not necessary to specify the included angle.

Slotted holes and internal chamfers present additional dimensioning situations and these are shown in Figure 7-14. Three methods are used for slotted holes; the choice depends on the function. Internal chamfers are dimensioned as shown at (b), with the special case shown by the 45° angle and the general case shown by the 90° angle. If dimensional control is required, two linear dimensions are used.

Irregular outlines are dimensioned as shown in Figure 7-15a or the *x-y* coordinates of the points can be tabulated. At (b) is shown the method of dimensioning symmetrical outlines. In this case the

(a)   (b)

Figure 7-14
Slotted holes for internal chambers.

(a)   (b)

Figure 7-15
Irregular and symmetrical outlines.

right side is shown and the left side is omitted. Note the break line and the symmetry symbol (=).

DIMENSIONING FOR LOCATION

After the geometric features that describe the shape of an object have been dimensioned, the location of these features is given. In some cases the size dimension will also double as the location dimension. However, in many situations, especially for cylindrical surfaces, another dimension is necessary.

Figure 7-16a illustrates one method of indicating the location dimensions for a hole. In this case, the hole on the left is located from two finished surfaces, the middle hole is located from the left hole, and the hole on the right is located from the middle hole. Dimensions are given to the center of the hole in both directions. This illustrates only the location of the holes and not the completely dimensioned part. Avoid dimensioning from unfinished surfaces, because unfinished surfaces cannot be relied upon for accurate measurements. Holes are finished surfaces, even though the finish mark is not used, and their centerlines can be used as extension lines.

Another method is illustrated at Figure 7-16b. It is referred to as rectangular coordinate dimension-

(a)   (b)

Figure 7-16
Hole location.

ing, and all dimensions are from common baselines which join at the lower left corner of the figure. The method to use is based on the function of the part; however, the latter method is recommended. Cylindrical surfaces should be located in the view in which they appear in their circular form.

Another example of rectangular coordinate dimensioning is shown in Figure 7-17a. In this case

(a)   (b)

Figure 7-17
Rectangular coordinate dimensioning.

HOLE	FROM	X	Y	Z
A	X,Y	35	35	10
B1	X,Y	10	10	THRU
B2	X,Y	10	54	THRU
B3	X,Y	60	54	THRU
B4	X,Y	60	10	THRU
C1	X,Y	35	10	4.2
C2	X,Y	35	54	4.2

HOLE	DESC.	QTY
A	Ø16	1
B	Ø6.2	4
C	Ø3.8	2

Figure 7-18
Rectangular coordinate dimensioning,
tabular form.

we have a cylindrical object and the coordinates are from the center of the object rather than the lower left corner as shown in Figure 7-16b. Rectangular coordinate dimensioning where the values are shown on the extension lines and the dimension lines are not used is shown in Figure 7-17b. The baselines are indicated as zero coordinates. Note the method of dimensioning the holes.

Figure 7-18 shows rectangular coordinate dimensioning in tabular form. Note the use of X, Y, and Z to denote the coordinates. Hole size and quantity are also shown in tabular form. Also, the depth of the hole is given under the Z column of the table.

The location dimensions for holes on a bolt circle are illustrated in Figure 7-19. The size of the hole, the number of holes, and the spacing are placed in notes. In this case we have 6 holes 60° apart, as indicated by the (6 × 60°) note. If the holes are unequally spaced, an angular dimension can be used to show their position. The bolt circle is always dimensioned in the view where the circle and the holes appear in circular form. This is the view that operating personnel use to lay off and scribe the center line of the holes. If the accuracy of the hole location is critical, a system of coordinate

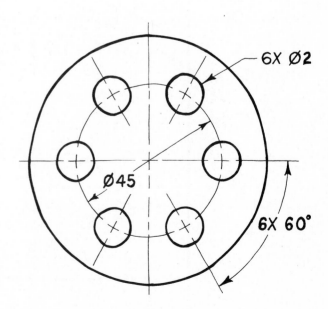

Figure 7-19
Bolt circle dimensioning.

dimensions (similar to those shown in Figure 7-17a) is used on each hole.

Proper dimensioning of a radial groove is illustrated in Figure 7-20. The groove is produced by an 18-mm-diameter milling cutter; therefore, the width of the groove is given as the diameter of the cutter. Radii at the ends of the groove are not necessary as these are the radius of the cutter. The center radius of the groove (the path of travel of the cutter) and angles locating the end extremities of the cutter are dimensional. To give additional dimensions merely makes it more difficult to machine and harder to check for accuracy. The figure further illustrates the method of dimensioning by polar coordinates, where the radii at the ends of the groove are dimensioned by their angular values (30° and 75°) and a linear value (114).

An example of a detailed part showing size and location dimensions is given in Figure 7-21. It should be studied to appreciate the difference between size, S, and location, L, dimensions. Location dimensions locate a part feature, and size dimensions define the physical size of the part feature. In some cases a dimension may be both a size and a location.

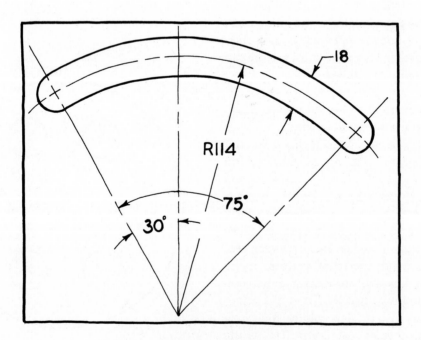

Figure 7-20
Radial groove dimensioning.

Figure 7-21
Size and location dimensions.

ADDITIONAL COMMENTS

While it is important that the views of the object be fully dimensioned, it is equally important that superfluous dimensions be eliminated. Extra dimensions can confuse operating personnel.

Another important consideration is the function of the part. Formerly, the manufacturing process was considered first and the function second. The present method is to tentatively dimension for the functional aspects, and then review the processes to see if any improvement can be made without adversely affecting the functional relationship.

The dimensioning conventions given in this chapter should be used as a guide. Some judgment is usually necessary when dimensioning the views of an object. A general review of most conventions is illustrated in Figure 7-22. Notice the placement of dimensions in the principal view and in the view where operating personnel will use them. Hole notes are in the circular view of the holes. Long extension lines and crowding of dimensions and notes

Figure 7-22
Dimensioning example.

have been avoided. As shown here, all dimensions and notes are placed in an open, easy-to-read space.

This chapter has included the major conventions. The reader is referred to the standard for information on knurling, keyseats, chords, and limited areas.

SUMMARY

In summary, one must know or do the following:

1. Operating personnel should not have to make calculations or scale a sketch.

2. Dimensioning should be accurate, clear, complete, and readable.

3. The first dimension line should be spaced 10 mm from the view and subsequent ones spaced 6 mm apart.

4. Special techniques are necessary for crowded spaces.

5. Larger dimensions should be outside the smaller dimensions.

6. Visible lines should not be used as extension lines.

7. A centerline may be used as an extension line.

8. Extension lines can cross all types of lines except dimension lines.

9. Visible and hidden lines must stand out from the other lines.

10. Dimensioning should be done off the views of the object when possible.

11. Dimensioning to hidden lines is to be avoided.

12. Leaders are required for local notes.

13. Finish marks are placed on all views of an object where the surface appears as an edge.

14. Finish marks are omitted on features where a note specifies a machining operation, a hole, or on surfaces of finished stock.

15. Dimensions should be given to finished surfaces or hole centers.

16. Dimensions are read from the bottom (unidirectional).

17. In the metric system, a zero precedes a value that is less than 1 (0.3).

18. In the metric system, the last digit to the right of the decimal period is not followed by a zero unless the value is part of a tolerance specification (3.5).

19. In the decimal inch system, a zero does not precede a value that is less than 1 (.5).

20. In the decimal inch system, zeros are added to the right to indicate accuracy (3.500).

21. The width, height, and depth of geometric features should be specified.

22. Cylinders should be dimensioned in their rectangular view.

23. Angles may be dimensioned by degrees and one coordinated dimension or by two coordinate dimensions, depending on the accuracy needed.

24. Reference values are enclosed by parentheses.

25. Holes are dimensioned by leaders.

26. The symbol Ø precedes a diameter value.

27. A small cross is placed at the radius center to indicate dimensional accuracy of the center.

28. When accuracy is desired, the location of holes in a circle is given by coordinate dimensions.

29. A chamfer is dimensioned by two linear values or one linear value and one angular value.

30. Cylindrical surfaces should be located in their circular view.

31. Rectangular coordinate dimensioning is the preferred method for locating holes and similar features.

32. Repetitive linear and angular dimensioning can effectively use the times (×) feature.

33. Polar coordinate dimensioning uses one linear value and one angular value.

34. Symbols are as follows:
S   Sphere
R   Radius
Ø   Diameter
⊔   Counterbore, Spotface
∨   Countersunk, Counterdrill
✓   Finish work
↧   Depth
×   Times/by

35. Superfluous dimensions should be eliminated.

36. In choosing dimensions, function is considered first and processes second.

37. Judgment is necessary to dimension an object.

TOLERANCING

Upon completion of this chapter, the student is expected to:

- Know the definition of tolerance and the techniques of expression.

- Understand the concept of tolerance accumulation.

- Understand the fundamentals of geometric tolerancing and the techniques of expression.

- Understand tolerance of form and the techniques of expression.

- Understand tolerance of profile and the techniques of expression.

- Understand tolerance of location and the techniques of expression.

- Understand tolerance of orientation and the techniques of expression.

- Understand tolerance of runout and the techniques of expression.

- Know the nomenclature of mating parts.

- Know the three types of fits for mating parts.

- Know the concept of the shaft basis and hole basis systems.

- Be able to determine the preferred sizes and fits using the metric system.

- Know the concept of surface texture and the techniques of expression.

INTRODUCTION

It is a truism in manufacturing that no two parts can be made exactly alike. There is a certain amount of variation in all the pieces that are produced of a particular part. The sources of this variation are machines, materials, environmental factors, and operating personnel.

The industrial revolution started the practice of interchangeable manufacturing wherein any part could be assembled into any device provided the part was within the prescribed limits. These prescribed limits are determined by the function of the part. For example, a tricycle would not need as tight limits as a watch or jet engine. The tighter the limits are, the more costly the manufacturing operation. Therefore, limits should be specified as loosely as possible provided the function of the device is not compromised.

In this chapter the metric system is used rather than the English system. The concepts are the same—only the numerical values are different. ANSI Y14.5M 1982 is the basis for this chapter.

TOLERANCING PRINCIPLES

Tolerance is the total amount a specific dimension is permitted to vary. It is the difference between the maximum and the minimum limits. In general there are three basic ways of expressing tolerance: direct dimensional tolerancing, geometric tolerancing, and notes.

Geometric tolerancing is discussed in much greater detail in later sections of this chapter. Notes can be quite effective in simplifying the tolerancing effort. A typical note might be UNLESS OTHERWISE SPECIFIED, ALL DIMENSIONS PLUS AND MINUS 0.2. A discussion of direct dimensional tolerancing follows.

Direct dimensional tolerancing takes on two forms: limit dimensioning and plus and minus tolerancing. **Limit dimensioning** is illustrated in Figure 8-1, and the tolerance is expressed by giving the limit dimensions with the high limit placed above the low limit. If the tolerance is expressed on a single line, the low limit is given first and the high limit next with a dash between the two. The same number of decimal places should be used for both limits. Toler-

Figure 8-1
Limit dimensioning.

ance for angles can be expressed in degrees, minutes, and seconds or by decimal fractions of a degree.

Plus and minus tolerancing is given in Figure 8-2. At (a) the concept of unilateral tolerancing is presented. Only one zero without a plus or minus sign is used. The basic value does not need to have the same number of decimal places as the tolerance. At (b) the concept of bilateral tolerancing is presented. Both the plus and minus values have the same number of decimal places.

All limits regardless of the number of decimal places are used as if they were continued with zeros. For example, 2.3 is equal to 2.300. . .0. A part feature is considered to be in nonconformance if it is outside the limits.

Single limits can be used where the intent is clear, in which case the letters MIN or MAX are placed after the dimension.

When more than one dimension and tolerance occur, a problem is created due to the combining of tolerances. Figure 8-3 shows a comparison of the tolerance accumulation of three different methods of dimensioning. At (a), the **chain dimensioning** method gives the maximum tolerance accumulation

(a) Unilateral tolerancing

(b) Bilateral tolerancing

Figure 8-2
Plus and minus tolerancing.

between surface b and c of ±0.3 or a total of 0.6 mm. The second technique is called **baseline dimensioning** and is illustrated at (b). In this case the tolerance accumulation between surface b and c is ±0.2 or a total of 0.4 mm. The third technique is called **direct dimensioning** and is illustrated at (c). In this case, the tolerance between surface b and c is 0.1 or a total of 0.2 mm, which is the lowest of the three techniques.

In certain cases it is necessary to indicate that a dimension between two features originates at one of the features and not the other. This technique is shown by a small circle, as at the bottom of Figure 8-3c.

When there are two or more adjacent dimensions with tolerances, it is important to consider the effect of one tolerance on another in relation to the desired function of the part. The dimensioning technique that is used can be instrumental in preventing an undesirable tolerance accumulation.

Different manufacturing processes are capable of different degrees of accuracy. Table 8-1 illustrates the approximate tolerance range for some shop processes with a basic size of 20 mm. It is also true

(a) Chain

(b) Baseline

(c) Direct

Figure 8-3
Tolerance accumulation.

TABLE 8-1
Tolerance for Various Processes at 20 mm

PROCESS	TOLERANCE RANGE (mm)
Lapping and honing	0.006 to 0.009
Grinding	0.009 to 0.021
Reaming	0.013 to 0.084
Turning	0.021 to 0.130
Drilling	0.084 to 0.130

that the larger the size of the part feature, the larger the tolerance. For example, a turning operation for a 100-mm-diameter feature can have a tolerance between 0.035 and 0.220, which is larger than the tolerance given in the table for a 20-mm diameter. The information in Table 8-1 is adapted from ANSI B4.2 and the designer can use that standard to assist in determining tolerances.

FUNDAMENTALS OF GEOMETRIC TOLERANCING

In simple terms, **geometric tolerancing** is described as a means of specifying the geometry of a part feature. It gives the designer a clear method of expressing the design intent, which in turn enables the manufacturer to choose the most appropriate method to produce the part. In addition, geometric tolerancing determines how the feature is to be inspected and gaged.

A part **feature** is a general term used to describe a physical portion of a part such as a surface, hole, or slot. On the other hand, a **feature of size** is a cylindrical or spherical surface or a set of parallel surfaces that is associated with a size dimension. Size dimensions were discussed in Chapter 7, as were location dimensions, which locate the centerline or centerplane of a part feature with another part feature, centerline, or centerplane.

To simplify the presentation of geometric tolerancing on a sketch, the following techniques are used:

1. Geometric characteristic symbols

2. Datum

3. Basic dimensioning

4. Material condition modifiers

5. Feature control frame

Geometric characteristic symbols are shown in Figure 8-4. Thirteen symbols are used and they are divided into groups. The *tolerance of form* group has the characteristics of straightness, flatness, circularity, and cylindricity. This group is used for individual features. The *tolerance of profile* group has the characteristics of profile of a line and profile of a surface. This group is used with either individual features or related features. The *tolerance of orientation* group has the characteristics of angularity, perpendicularity, and parallelism; the *tolerance of location* group has the characteristics of position and concentricity; and the *tolerance of runout* has the characteristics of circular runout and total runout. These three groups are used with two or more related features. The five groups are discussed in greater detail later in the chapter.

A **datum** indicates the origin of a dimensional relationship between a toleranced feature and designated feature(s) on a part. Letters are used to specify a datum. Figure 8-5 shows two datums, M and K, for a cylindrical surface and a plane surface, respectively.

	TYPE OF TOLERANCE	CHARACTERISTIC	SYMBOL
FOR INDIVIDUAL FEATURES	FORM	STRAIGHTNESS	—
		FLATNESS	▱
		CIRCULARITY (ROUNDNESS)	○
		CYLINDRICITY	⌒⁄
FOR INDIVIDUAL OR RELATED FEATURES	PROFILE	PROFILE OF A LINE	⌒
		PROFILE OF A SURFACE	⌓
FOR RELATED FEATURES	ORIENTATION	ANGULARITY	∠
		PERPENDICULARITY	⊥
		PARALLELISM	//
	LOCATION	POSITION	⊕
		CONCENTRICITY	◎
	RUNOUT	CIRCULAR RUNOUT	↗
		TOTAL RUNOUT	⟋⟋

Figure 8-4
Geometric characteristic symbols.

Figure 8-5
Datum and basic dimensioning.

117

(a) External feature

(b) Internal feature

Figure 8-6
MMC and LMC of features of size.

A **basic dimension** is a theoretical exact dimension without tolerance. It is a numerical value that is enclosed in a rectangle, as shown by the dimensions 40.0 and 15.0 in Figure 8-5.

There are three material condition modifiers: maximum material condition (MMC), least material condition (LMC), and regardless of feature size (RFS). These modifiers can be used only with features of size, such as holes, shafts, and slots, rather than with features such as surfaces.

When the feature of size contains the greatest amount of material it is at the **maximum material condition.** The MMC of an external feature of size (shaft) is the largest size and the MMC of an internal feature of size (hole) is the smallest size, as shown in Figure 8-6.

When the feature of size contains the least amount of material, it is at the **least material condition.** The LMC of an external feature of size (shaft) is the smallest size and the LMC of an internal feature of size (hole) is the largest size, as shown in Figure 8-6.

When the feature of size is not at either extreme, but at any size including MMC and LMC, it is referred to as **regardless of feature size.** The RFS occurs when a geometric tolerance is applied independent of the feature size. In other words the geometric tolerance is given as the stated amount for each feature of size.

A **feature control frame** specifies the type, shape, and size of the geometric tolerance zone. It also dictates the datum features and assigns the material condition modifiers to the tolerance and the datum when applicable. A typical feature control frame is shown in Figure 8-7.

Figure 8-7
Typical feature control frame.

THREE RULES OF TOLERANCING

In geometric tolerancing, three fundamental rules form the foundation of the system. **Rule 1** states that where only a tolerance of size is specified, the limits of size prescribe the amount of variation permitted in its form. Figure 8-8 illustrates this concept for the straightness characteristic using an external feature at the top and an internal feature at the bottom. The feature of size is shown at (a) and its meaning at (b). Note that when the actual size is at the MMC value, no variation in form is permitted. In other words, perfect form exists at MMC. Where the actual size of the part feature departs from the MMC value, a variation of form is allowed. This variation is equal to the amount of that departure. For a shaft with a tolerance value of Ø 25.4–25.8 the variation in form is as follows:

ACTUAL SIZE	VARIATION IN FORM
25.4 LMC	0.4
25.5	0.3
25.6	0.2
25.7	0.1
25.8 MMC (perfect)	0

(a) Tolerance　　　　　　(b) Interpretation

Figure 8-8
Straightness example for rule 1.

For a hole with a tolerance value of Ø 26.2–26.7, the variation in form is as follows:

ACTUAL SIZE	VARIATION IN FORM
26.2 MMC (perfect)	0
26.3	0.1
26.4	0.2
26.5	0.3
26.6	0.4
26.7 LMC	0.5

Note that Figure 8-8 only shows rule 1 for straightness. Similar interpretations for flatness, circularity, and cylindricity are left as exercises. Also, rule 1 does not apply to stock sizes such as bars, sheets, tubing, and so on, or to nonrigid parts such as some plastics.

Rule 2 states that where a tolerance of position is specified, RFS (symbolized by S), LMC (symbolized by L), or MMC (symbolized by M) must also be specified with respect to the tolerance, the datum, or both. The feature control frame that illustrates rule 2 is shown in Figure 8-9a.

Rule 3 states that for other than tolerance of position (rule 2), RFS applies with respect to the tolerance, the datum, or both, unless MMC is specified. Furthermore, it is not necessary to include the RFS symbol in the feature control frame. The rule is illustrated in Figure 8-9b.

(a) Rule #2

(b) Rule #3

Figure 8-9
Feature control frame examples
of rules 2 and 3.

DATUM REFERENCING

Datum planes are associated with process or inspection equipment such as machine tables, surface plates, collets, and locating pins. Although these planes are not true planes, they are of such quality that they are used to simulate the true planes, from which measurements are made to verify dimensions and tolerances. A **datum feature** is the inexact part feature that corresponds to the datum plane.

To describe a part accurately, three mutually perpendicular planes are required. These three planes are shown in Figure 8-10 and are called the **datum reference frame.** A part is placed in contact with the primary datum, where three points are needed to make contact with the datum. Two points of contact are needed for the secondary datum, and one point is needed for the tertiary datum. The primary datum plane is listed first, the secondary datum next, and the tertiary datum last in the feature control frame; however, depending on need, only the primary datum or primary and secondary datums may be necessary. Datum features and datum symbols that are related to the datum planes are shown by a two-view sketch at (b).

For cylindrical parts, the three-plane reference frame is somewhat more difficult to visualize. The

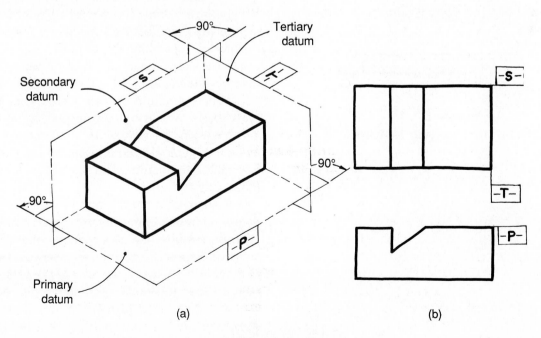

(a)   (b)

Figure 8-10
Datum reference frame.

Figure 8-11
Cylindrical datum features.

primary datum, D, is the flat surface perpendicular to axis of the cylindrical datum feature as shown in Figure 8-11. Datum E is associated with two theoretical planes at 90° to each other that intersect at the axis of the cylinder. These two planes are represented on the sketch by the centerlines of the cylinder and are used to indicate the direction of measurement in an x and y direction. Note that the datum symbol is placed with the cylinder limits.

There are two types of datum features: surface and size. A datum surface has no size tolerance because it is a plane from which dimensions or relationships originate. Where a normally flat surface is specified as a datum feature, the corresponding datum is simulated by a plane contacting the high points of that surface. The extent of contact depends on whether the surface is a primary, secondary, or tertiary datum feature.

On the other hand, datums of size are established from features of size with tolerances such as slots and holes. Because variations are allowed by the size dimension, it is necessary to determine whether RFS or MMC will be used. For a tolerance of position, the datum reference letter is always followed by the material condition modifier in the feature control frame (rule 2); for all other geometric tolerances, RFS is implied unless otherwise specified (rule 3).

Where a datum feature of size is applied on a RFS basis, the datum is established by physical contact between the feature surface and the surface of the processing equipment. For example, a chuck on a lathe makes contact at different part diameters, thereby establishing a RFS datum for each part.

Where a datum feature of size is applied on a MMC basis, processing and inspection equipment, which remain constant in size, may be used to simulate a true geometric counterpart of the feature and to establish the datum. The size of the datum is the MMC value.

Instead of using a plane to locate a part in a datum reference frame, datum targets can be used. **Datum targets** are designated points, lines, or areas of contact. They are described on the sketch of the part; however, they refer to the gage features that are used to simulate the datum planes.

A datum target **point** is indicated on the part sketch with an X as shown in Figure 8-12. The point contact is made possible by spherical or pointed locating pins. Datum target symbols are placed outside the part outline with a leader to the target. A

Figure 8-12
Datum target point.

solid leader indicates that the target is visible and a dashed leader indicates that the target is invisible, which is the case in the figure. The symbol is a circle with a horizontal line creating an upper and a lower half. A letter and a numeral are placed in the lower half to identify the datum (C) and the point (1) on the datum. Information for the upper half of the circle is given later in this section. The datum target point is usually located by basic dimensioning with gage tolerances assumed to apply.

Datum target lines are used whenever points are unable to support a part adequately. Also, points might damage a part. However, lines are not as accurate in locating a part as points. An X and a phantom line are used to indicate a target line, as shown in Figure 8-13.

Figure 8-13
Datum target line.

Figure 8-14
Datum target area.

Datum target **areas** are used where additional support for the part is necessary and accuracy is not critical. The area is specified with a crosshatched circle surrounded by a phantom line or an X if a circle is impractical. As shown in Figure 8-14, the diameter of the area is given in the upper half of the symbol. The figure illustrates the use of the areas of three locating pins to simulate the true or theoretical datum plane A.

TOLERANCE OF FORM

Form tolerances control flatness, straightness, circularity, and cylindricity. These tolerances are specified where the tolerances of size or location do not provide sufficient control. They are applicable to individual (single) features and are, therefore, not related to datums.

Flatness is the condition of a surface having all elements in one plane. A flatness tolerance specifies a tolerance zone defined by two parallel planes within which the entire surface must lie, as shown in Figure 8-15. The feature control frame is attached to an extension line of the surface or to a leader with an arrowhead pointing to the applicable surface. It is placed in a view where the surface is represented by a line.

Straightness is the condition in which an *element* of a surface or an axis is a straight line. A straightness tolerance specifies a tolerance zone within which the considered element of the surface or axis of a cylinder must lie. It differs from flatness in that flatness covers an entire surface rather than an element of the surface. When straightness is

(a) Tolerance

(b) Interpretation

Figure 8-15
Flatness tolerance.

(a) Tolerance

(b) Interpretation

Figure 8-16
Straightness tolerance
for a cylindrical surface.

specified for a surface, the feature control frame is attached to an extension of a surface element or to a leader with an arrowhead pointing to a surface element. Figure 8-16 illustrates the concept for a cylindrical surface. Each longitudinal element of the circular surface must lie within two parallel lines that are parallel to the axis of the cylinder. The straightness tolerance must be within the limits of size tolerance and the boundary of perfect form at MMC. Figure 8-17 illustrates the concept for a flat surface. Straightness may be applied to line elements in a single direction or to line elements in two directions as shown in the figure. Each longitudinal element must lie between two parallel lines 0.06 apart in the front view and 0.09 in the side view.

Where straightness is specified for an axis, the feature control frame is placed with the feature size

(a) Tolerance

(b) Interpretation

Figure 8-17
Straightness tolerance
for a flat surface.

dimension. If a cylindrical feature is being specified, the diameter symbol is required to indicate a cylindrical tolerance zone. The foregoing information is shown in Figure 8-18. It was stated previously that rule 1 controls the four tolerances of form. However, when the axis straightness is applied to a shaft or hole, the feature is no longer required to meet the perfect form at MMC. The absence of a modifier assumes RFS. When the part feature is at 18.2, dimension x can be 18.5, and when the part feature is at 18.8, dimension x can be 19.1, which is the maximum value and is called the **virtual condition.** *Note:* If MMC had been specified, the diameter tolerance zone would have increased as the part feature departed from the MMC value and would equal 0.9 at LMC.

Circularity is a condition of a surface of revolution (cylinder, cone, or sphere) where all points of the surface intersected by any plane perpendicular to a common axis (cylinder or cone) or passing through a common center (sphere) are equidistant from the center. A circularity tolerance specifies a tolerance zone bounded by two concentric *circles* within which the circular element must lie. Figure 8-19 illustrates the concept. The tolerance applies independently at any plane along the feature from point a to point b.

Cylindricity is a condition of a surface of revolution in which all points of the surface are equidistant from a common axis. A cylindricity tolerance specifies a tolerance zone bounded by two concentric *cylinders* within which the surface must lie. The tolerance callout and interpretation are shown in

(a) Tolerance (b) Interpretation

**Figure 8-18
Straightness tolerance
for an axis.**

(a) Tolerance   (b) Interpretation

Figure 8-19
Circularity tolerance.

(a) Tolerance   (b) Interpetation

Figure 8-20
Cylindricity tolerance.

Figure 8-20. Cylindricity tolerance is a combination of circularity and straightness applied to a cylindrical object.

TOLERANCE OF PROFILE

Profile tolerancing is a method of controlling irregular surfaces, lines, arcs, or unusual shapes, as well as regular shapes. Profiles can be applied to individual line elements called **profile of a line** or to the entire surface called **profile of a surface** of a part. The difference between profile of a line and profile of a surface is analogous to the difference between circularity and cylindricity. Profile tolerancing is unique in that it can be applied to individual features (no datum) or to related features (with a datum).

A profile tolerance specifies a tolerance zone along the true profile within which the elements of

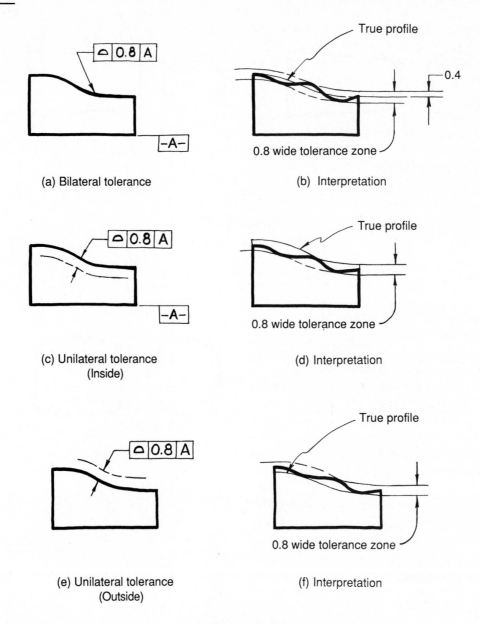

Figure 8-21
Profile tolerance of a surface.

the surface or the entire surface must lie. The profile of a surface tolerance is three-dimensional and may be applied to surfaces having a constant cross section, whereas the profile of a line is two-dimensional and applies to parts having a varying cross section, such as the tapered wing of an airplane. The tolerance zone is the distance between two boundaries shaped to the true profile that has been defined by basic dimensions. Two different methods of specifying the tolerance zone can be used as shown in Figure 8-21. In the bilateral method the zone is divided on both sides of the true profile and the boundaries do not need to be shown. However, in the unilateral method a boundary (phantom line) is needed to show the direction of the zone from the true profile.

In most cases a datum reference is necessary to provide proper orientation for a profile of a surface. For a profile of a line, a datum is usually not needed. If it is necessary to indicate the extent of the profile, that information is given below the feature control frame using ALL AROUND or BETWEEN H&L. It is also possible to specify a different tolerance for different portions of a surface or line. A profile tolerance can be combined with other types of geometric tolerancing.

TOLERANCE OF ORIENTATION

Angularity, perpendicularity, and parallelism are the three orientation tolerances. They control the orientation of a feature or features to a datum feature. All three can be applied to a feature or a feature of size and can use a material condition modifier. When they are applied to plane surfaces, the flatness of the surface is controlled to within the orientation tolerance.

Angularity is the condition of a surface or axis at a specified angle (other than 90°) from a datum plane or axis. An angularity tolerance specifies a tolerance zone defined by two parallel planes at a specific angle with the datum. All points on the angular surface or angular axis must lie between these two planes. Figure 8-22 illustrates the concept for a sur-

(a) Tolerance   (b) Interpretation

**Figure 8-22
Angularity tolerance of a surface.**

Figure 8-23
Perpendicularity tolerance of
a surface.

face. Note that the angle is enclosed in the basic box that is required for an angularity tolerance.

Perpendicularity is the condition of a surface, center plane, or axis at a right angle to a datum plane or axis. A perpendicularity tolerance specifies a tolerance zone defined by two parallel planes, two parallel lines, or a cylinder at 90° to a datum. All elements of the surface, center plane, or axis must fall within the tolerance. Figure 8-23 illustrates the concept for a perpendicular surface.

When the perpendicularity tolerance zone is a cylinder, the diameter symbol must be included in the feature control frame as shown in Figure 8-24. This figure also illustrates the application of the tolerance to a feature of size. In this case the diameter

Figure 8-24
Perpendicularity tolerance of
an axis.

(a) Tolerance

(b) Interpretation

Figure 8-25
Parallelism tolerance of surface.

of the cylindrical tolerance zone remains at 0.2 since RFS is assumed. However, if MMC had been specified after the 0.2 value in the feature control frame, the tolerance zone diameter would vary from 0.2 at MMC to as much as 0.5 at LMC. Thus, as the feature of size departs from the MMC value, a bonus tolerance occurs.

Parallelism is the condition of a surface equidistant at all points from a datum plane or an axis equidistant along its length to a datum axis. A parallelism tolerance specifies a tolerance zone defined by two parallel planes, two parallel lines, or a cylinder that is parallel to a datum. Figure 8-25 illustrates the concept for a surface.

TOLERANCE OF LOCATION

Position, concentricity, and symmetry are the three tolerances of location. Location controls deal with features of size only and require one of the three material condition modifiers to be specified in the feature control frame. Also, locational controls require at least one datum reference.

Positional tolerance defines a zone in which the center, axis, or center plane is permitted to vary from the exact (perfect) location, which is referred to as its *true position*. A positional tolerance is the total permissible variation in the location of the feature about its true position. The positional tolerance is indicated by its symbol, tolerance, modifier, and datum reference in the feature control frame. For cylindrical features, the tolerance is generally the diameter of the tolerance zone that contains the axis

(a) Tolerance   (b) Interpetation   (c) Square tolerance zone

Figure 8-26
Positional tolerance.

(true position) of the feature of size. For features that are not round, such as slots and tabs, the positional tolerance is the total width of the tolerance zone in which the center plane (true position) of the feature must lie.

Figure 8-26a shows the tolerance with basic dimensions establishing the true position of the axis of the hole. At (b) is shown the interpretation of the tolerance. When limit or plus and minus dimensions are used to locate the axis, a square tolerance zone occurs as shown at (c). This latter technique is not recommended because most fixed pin gages use round pins, which cannot evaluate a square tolerance zone. In addition, tolerance accumulation occurs when a second feature of size is located from the first.

A positional tolerance applied at MMC may be explained in terms of the surface of a hole or in terms of the axis of a hole. For a surface, the hole position may vary, but no element of its surface can be inside the theoretical boundary. As shown in Figure 8-27, the theoretical boundary is the MMC minus the positional tolerance.

Where a hole is at MMC, its axis must fall within a cylindrical tolerance zone whose axis is located at the true position. The diameter of this zone is equal to the positional tolerance. Figure 8-28a shows the tolerance with its interpretation shown at (b). Three holes are used to illustrate the concept. The hole at the left has the axis centered in the cylin-

Figure 8-27
Boundary for a surface
of a hole at MMC.

(a) Tolerance

(b) Interpretation

Figure 8-28
Hole axis in relation to
positional tolerance zones.

(a) Tolerance

(b) Interpretation

Figure 8-29
Projected tolerance zone.

(a) Tolerance

(b) Interpretation

Figure 8-30
Concentricity tolerance.

drical tolerance zone; the hole in the middle has the axis on the edge of the zone; and the hole at the right has the axis on an angle. In all three cases the axis is inside the tolerance zone. It is only when the feature is at MMC that the specified tolerance applies. As the feature of size departs from the MMC, an additional positional tolerance results. For example, where the feature of size is 15.5, the positional tolerance is Ø 0.4. Similarly, for features of size of 15.6 and 15.7, the positional tolerances are Ø 0.5 and 0.6, respectively.

Occasionally, there is a need to extend a tolerance zone beyond the surface of the part. This action can be indicated by using the projected tolerance symbol, P, and specifying the height of the theoretical tolerance zone above the surface. Figure 8-29 illustrates the concept. Note that the symbol and height of the zone are given in a box below the feature control frame.

In certain cases, the design or function of a part may require the positional tolerance, datum reference, or both to use an RFS material condition modifier. Since each feature of size axis must be located within the specified tolerance zone, the requirement imposes a closer control of the features. Also, verification of the specification is quite intricate.

Where positional tolerancing at LMC is used, the stated positional tolerance applies to its LMC size. Perfect form is required at LMC. Where the feature departs from its stated LMC size, an increase in the positional tolerance is allowed and is equal to the amount of the departure. This additional tolerance is the same as discussed previously under MMC. Specifying LMC is not too common unless considerations critical to the design are required.

Concentricity is the condition where the axis of all cross-sectional elements of a surface of revolution (feature of size) are common to the axis of a datum reference. A concentricity tolerance specifies a cylindrical tolerance zone whose axis coincides with a datum axis (usually the larger diameter) and within which the axis of the considered feature (smaller diameter) must lie. Figure 8-30 illustrates the concept. A concentricity tolerance is established only on an RFS basis. Verification of the tolerance is difficult; therefore, concentricity specifications should be avoided where possible.

Symmetry is a condition in which a feature or features are symmetrically disposed about the center plane of a datum feature. The positional tolerance symbol is used to specify symmetry. A symmetrical tolerance is specified by two parallel planes within which the center plane of the feature must lie. Figure 8-31 illustrates the concept for a slot. In this case the center plane of the slot must lie between two parallel planes 0.8 apart, regardless of the feature size, which are equally disposed about the center plane of datum B.

(a) Tolerance   (b) Interpretation

Figure 8-31
Symmetry tolerance.

TOLERANCE OF RUNOUT

Circular runout and total runout are the two tolerances of runout. Runout is combination control affecting both the form (surface) and location (position) of a part feature to a datum axis. The types of surfaces controlled by the runout tolerances are those surfaces constructed around the datum axis and those surfaces perpendicular to the datum axis as shown in Figure 8-32. Runout control requires a datum reference and the material condition modifier is RFS. A runout tolerance value is the full indicator movement, FIM, of the considered feature when the part is rotated 360° about its datum axis. Runout is used most frequently on objects that have circular cross sections resulting from turning operations such as would occur on a lathe.

Circular runout provides control of circular *elements* of a surface. Figure 8-32 shows the application of the tolerance to both a surface constructed around a datum axis and one perpendicular to a datum axis. The tolerance is applied independently to any circular element.

Total runout provides combination control of all surface elements. The tolerance is applied simultaneously to circular and longitudinal elements as the part is rotated 360°. Figure 8-33 shows the application of the tolerance to both a surface constructed around a datum axis and one perpendi-

(a) Tolerance   (b) Interpretation

Figure 8-32
Circular runout tolerance.

(a) Tolerance   (b) Interpretation

Figure 8-33
Total runout tolerance.

cular to a datum axis. Where it is applied to surfaces around the axis, it controls the cumulative variations of circularity, cylindricity, straightness, coaxiality, angularity, taper, and profile. Where it is applied to surfaces perpendicular to the axis, it controls cumulative variations of perpendicularity and flatness.

LIMITS AND FITS
FOR MATING PARTS

Whenever two parts are designed to function together, they are mating parts. One of the mating parts is classified as external and the other, internal. The shaft and hole is the most common external-internal pair; however, the reader should note that there are other pairs, such as key-keyway, T-bolt-slot, and electrical plug-outlet. Our discussion will use the shaft-hole pair to illustrate the concept. The basis of the material in this section is ANSI B4.2-1978, Preferred Metric Limits and Fits. A similar system is used for the English system.

To use the system, an understanding of the nomenclature is required. Figure 8-34 shows a shaft-hole pair and the appropriate terms. For simplicity the shaft and the hole are sketched coincident to each other on the left side. The maximum and minimum allowable size of the shaft and of the hole are shown on the figure. As stated previously, the

Figure 8-34
Nomenclature for mating parts.

difference in the maximum and minimum values is the tolerance.

The **basic size** is a value from which the deviations are applied to determine the maximum and minimum sizes of the parts. These values should be selected from Table 8-2. Whenever possible, you should select basic sizes from the first column because they represent standard sizes for materials and tools, thereby reducing cost. Also, the tables are based on these preferred sizes.

The **upper deviation** is the difference between the maximum permissible size of a part and the basic size. For a shaft this difference is the largest shaft (MMC) minus the basic size; for a hole this difference is the largest hole (LMC) minus the basic size. The **lower deviation** is the difference between the minimum permissible size of a part and the basic size. For a shaft, this difference is the smallest shaft (LMC) minus the basic size; for a hole, this difference is the smallest hole (MMC) minus the basic size. That deviation which is closest to the basic size is called the **fundamental deviation.** For the shaft, it is the upper deviation; for the hole, it is the lower deviation and is specified with an uppercase letter such as H. Note that in both cases, the fundamental deviation involves the MMC.

An **international tolerance (IT) grade** is used to specify the tolerance. There are 18 IT grades, ranging from IT01, IT0, IT1, to IT16. Within each grade, the tolerances vary in accordance with the basic size. This system provides a uniform level of accuracy.

The **tolerance zone** is a combination of the fundamental deviation and the tolerance grade as shown in Figure 8-35. For a shaft, the basic size is first, then the shaft fundamental deviation (lowercase letter), and then the IT grade; for a hole the basic size is first, then the hole fundamental deviation (uppercase letter), and then the IT grade. The fit is specified by the basic size, hole tolerance zone, and then the shaft tolerance zone. Four methods of showing the tolerance are given in Figure 8-36. Where information is given in parentheses, it means for reference only.

TABLE 8-2
Preferred Sizes

BASIC SIZE (MM)		BASIC SIZE (MM)		BASIC SIZE (MM)	
FIRST CHOICE	SECOND CHOICE	FIRST CHOICE	SECOND CHOICE	FIRST CHOICE	SECOND CHOICE
1		10		100	
	1.1		11		110
1.2		12		120	
	1.4		14		140
1.6		16		160	
	1.8		18		180
2		20		200	
	2.2		22		220
2.5		25		250	
	2.8		28		280
3		30		300	
	3.5		35		350
4		40		400	
	4.5		45		450
5		50		500	
	5.5		55		550
6		60		600	
	7		70		700
8		80		800	
	9		90		900
				1000	

Figure 8-35
Tolerance symbols for mating parts.

$$30\text{H}7 \qquad 30\text{H}7\binom{30.021}{30.000} \qquad \frac{30.021}{30.000} \qquad \frac{30.021}{30.000}\left(30\text{H}7\right)$$

(a)   (b)   (c)   (d)

Figure 8-36
Methods for specifying tolerance.

There are three types of fits: clearance, interference, and transition. A **clearance fit** occurs where there is *always* an air space between two mating parts regardless of the part sizes (provided that they are within tolerance). As can be seen in Figure 8-37a, the shaft is smaller than the hole under all tolerance conditions. An **interference fit** occurs where there is *always* an interference of material between two mating parts regardless of the part sizes (provided that they are within tolerance). As can be seen at (b), the shaft is larger than the hole under all tolerance conditions. The assembly of these two mating parts requires that they be forced together. A **transition fit** is one where *sometimes* there is an air space and *sometimes* there is material interference, depending on mating part sizes. The concept is illustrated at (c). When the shaft is at MMC (largest shaft) and the hole is at MMC (smallest hole), there is material interference; however, when both features are at LMC, there is an air space. Transition fits are used with location dimensions.

Preferred fits for clearance, transition, and interference are given in Table 8-3. Wherever possible

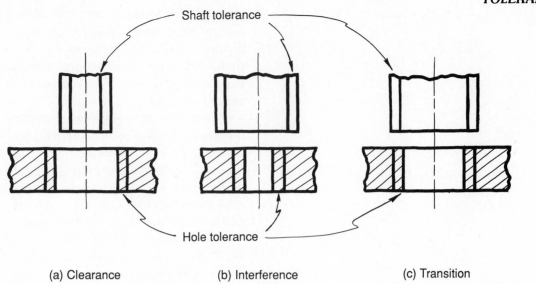

(a) Clearance   (b) Interference   (c) Transition

Figure 8-37
Hole tolerance.

TABLE 8-3
Description of Preferred Fits

HOLE BASIS	SHAFT BASIS	DESCRIPTION
		Clearance Fits
H11/c11	C11/h11	*Loose-running* fit for wide commercial tolerances or allowances on external members.
H9/d9	D9/h9	*Free-running* fit—not for use where accuracy is essential, but good for large temperature variations, high running speeds, or heavy journal pressures.
H8/f7	F8/h7	*Close-running* fit for running on accurate machines and for accurate location at moderate speeds and journal pressures.
H7/g6	G7/h6	*Sliding* fit—not intended to run freely, but to move and turn freely and locate accurately.
		Transition Fits
H7/h6	H7/h6	*Locational clearance* fit provides snug fit for locating stationary parts, but can be freely assembled and disassembled.
H7/k6	K7/h6	*Locational transition* fit for accurate location, a compromise between clearance and interference.
H7/n6	N7/h6	*Locational transition* fit for more accurate location where greater interference is permissible.
		Interference Fits
H7/p6[a]	P7/h6	*Locational interference* fit for parts requiring rigidity and alignment with prime accuracy of location but without special bore pressure requirements.
H7/s6	S7/h6	*Medium drive* fit for ordinary steel parts or shrink fits on light sections, the tightest fit usable with cast iron.
H7/u6	U7/h6	*Force* fit suitable for parts that can be highly stressed or for shrink fits where the heavy pressing forces required are impractical.

[a]Transition fit for basic sizes in range from 0 through 3 mm.

fits should be selected from this table. The description provides information for the designer to select the appropriate fit. Fits are given for hole basis and shaft basis. These two systems were developed to aid the practitioner in the determination of tolerances of mating parts.

The **hole basis system** uses the smallest hole (MMC) as the basic size; therefore, its lower deviation is 0. It is the most common system because standard reamers, broaches, and other tools are used to produce holes and standard plug gages are used to inspect them. Shafting, on the other hand, can be machined to any desired size.

A practical example will aid in understanding the system. The designer finds that a free-running fit will meet the criteria for two mating parts. Also, design calculations give a diameter of approximately 29 mm.

Tolerance symbols

From Table 8-3 obtain   H9/d9

Basic size

From Table 8-2 select the preferred size of Ø 30

Hole size (H9)

From Table 8-4 obtain   Ø 30.000–30.052

Shaft size (d9)

From Table 8-4 obtain   Ø 29.883–29.935

Tightest fit

Smallest hole (MMC) − largest shaft (MMC)
30.000 · 29.935 = 0.065

Loosest fit

Largest hole (LMC) − smallest shaft (LMC)
30.052 − 29.883 = 0.169

Note that the 0.065 and 0.169 values are given in the table; therefore, it is not necessary to make the calculations. They are given as an aid to understanding the concept.

The **shaft basis system** uses the largest shaft (MMC) as the basic size; therefore, its upper deviation is 0. While the shaft basis system is not as common, it does have applications in standard cold-finished shafting. A boring operation on the hole is needed to achieve its desired size.

A practical example will aid in understanding the system. The designer finds that a medium-drive fit will meet the criteria for two mating parts. Also, design calculations give a diameter of 15 mm.

Tolerance symbols

From Table 8-3 obtain   S7/h6

Basic size

From Table 8-2 select the preferred size of 16

Hole size (S7)

From Table 8-5 obtain   Ø 15.961-15.979

Shaft size (h6)

From Table 8-5 obtain   Ø 15.989-16.00

Tightest fit

Smallest hole (MMC) − largest shaft (MMC)
15.961 − 16.000 = −0.039

Loosest fit

Largest hole (LMC) − smallest shaft (LMC)
15.979 − 15.989 = 0.010

In the shaft basis system, the fits are also given in the table.

Before completing our discussion of mating parts, the concept of selective assembly should be explained. When precise parts (small tolerance) are needed for complicated devices to function properly, the demands on the processing equipment may be greater than their capability. When this occurs, there are two alternatives: (1) discard the parts that exceed the limits (expensive), or (2) use manual or computer-controlled selective assembly. In selective assembly, all parts are sorted and classified

TABLE 8-4
Preferred Hole Basis Fits for Diameters to 30 mm

BASIC SIZE		LOOSE RUNNING Hole H11	Shaft c11	Fit	FREE RUNNING Hole H9	Shaft d9	Fit	CLOSE RUNNING Hole H8	Shaft f7	Fit	SLIDING Hole H7	Shaft g6	Fit	LOCATIONAL CLEARANCE Hole H7	Shaft h6	Fit
1	MAX	1.060	0.940	0.180	1.025	0.980	0.070	1.014	0.994	0.030	1.010	0.998	0.018	1.010	1.000	0.016
	MIN	1.000	0.880	0.060	1.000	0.955	0.020	1.000	0.984	0.006	1.000	0.992	0.002	1.000	0.994	0.000
1.2	MAX	1.260	1.140	0.180	1.225	1.180	0.070	1.214	1.194	0.030	1.210	1.198	0.018	1.210	1.200	0.016
	MIN	1.200	1.080	0.060	1.200	1.155	0.020	1.200	1.184	0.006	1.200	1.192	0.002	1.200	1.194	0.000
1.6	MAX	1.660	1.540	0.180	1.625	1.580	0.070	1.614	1.594	0.030	1.610	1.598	0.018	1.610	1.600	0.016
	MIN	1.600	1.480	0.060	1.600	1.555	0.020	1.600	1.584	0.006	1.600	1.592	0.002	1.600	1.594	0.000
2	MAX	2.060	1.940	0.180	2.025	1.980	0.070	2.014	1.994	0.030	2.010	1.998	0.018	2.010	2.000	0.016
	MIN	2.000	1.880	0.060	2.000	1.955	0.020	2.000	1.984	0.006	2.000	1.992	0.002	2.000	1.994	0.000
2.5	MAX	2.560	2.440	0.180	2.525	2.480	0.070	2.514	2.494	0.030	2.510	2.498	0.018	2.510	2.500	0.016
	MIN	2.500	2.380	0.060	2.500	2.455	0.020	2.500	2.484	0.006	2.500	2.492	0.002	2.500	2.494	0.000
3	MAX	3.060	2.940	0.180	3.025	2.980	0.070	3.014	2.994	0.030	3.010	2.998	0.018	3.010	3.000	0.016
	MIN	3.000	2.880	0.060	3.000	2.955	0.020	3.000	2.984	0.006	3.000	2.992	0.002	3.000	2.994	0.000
4	MAX	4.075	3.930	0.220	4.030	3.970	0.090	4.018	3.990	0.040	4.012	3.996	0.024	4.012	4.000	0.020
	MIN	4.000	3.855	0.070	4.000	3.940	0.030	4.000	3.978	0.010	4.000	3.988	0.004	4.000	3.992	0.000
5	MAX	5.075	4.930	0.220	5.030	4.970	0.090	5.018	4.990	0.040	5.012	4.996	0.024	5.012	5.000	0.020
	MIN	5.000	4.855	0.070	5.000	4.940	0.030	5.000	4.978	0.010	5.000	4.988	0.004	5.000	4.992	0.000
6	MAX	6.075	5.930	0.220	6.030	5.970	0.090	6.018	5.990	0.040	6.012	5.996	0.024	6.012	6.000	0.020
	MIN	6.000	5.855	0.070	6.000	5.940	0.030	6.000	5.978	0.010	6.000	5.988	0.004	6.000	5.992	0.000
8	MAX	8.090	7.920	0.260	8.036	7.960	0.112	8.022	7.987	0.050	8.015	7.995	0.029	8.015	8.000	0.024
	MIN	8.000	7.830	0.080	8.000	7.924	0.040	8.000	7.972	0.013	8.000	7.986	0.005	8.000	7.991	0.000
10	MAX	10.090	9.920	0.260	10.036	9.960	0.112	10.022	9.987	0.050	10.015	9.995	0.029	10.015	10.000	0.024
	MIN	10.000	9.830	0.080	10.000	9.924	0.040	10.000	9.972	0.013	10.000	9.986	0.005	10.000	9.991	0.000
12	MAX	12.110	11.905	0.315	12.043	11.950	0.136	12.027	11.984	0.061	12.018	11.994	0.035	12.018	12.000	0.029
	MIN	12.000	11.795	0.095	12.000	11.907	0.050	12.000	11.966	0.016	12.000	11.983	0.006	12.000	11.989	0.000
16	MAX	16.110	15.905	0.315	16.043	15.950	0.136	16.027	15.984	0.061	16.018	15.994	0.035	16.018	16.000	0.029
	MIN	16.000	15.795	0.095	16.000	15.907	0.050	16.000	15.966	0.016	16.000	15.983	0.006	16.000	15.989	0.000
20	MAX	20.130	19.890	0.370	20.052	19.935	0.169	20.033	19.980	0.074	20.021	19.993	0.041	20.021	20.000	0.034
	MIN	20.000	19.760	0.110	20.000	19.883	0.065	20.000	19.959	0.020	20.000	19.980	0.007	20.000	19.987	0.000
25	MAX	25.130	24.890	0.370	25.052	24.935	0.169	25.033	24.980	0.074	25.021	24.993	0.041	25.021	25.000	0.034
	MIN	25.000	24.760	0.110	25.000	24.883	0.065	25.000	24.959	0.020	25.000	24.980	0.007	25.000	24.987	0.000
30	MAX	30.130	29.890	0.370	30.052	29.935	0.169	30.033	29.980	0.074	30.021	29.993	0.041	30.021	30.000	0.034
	MIN	30.000	29.760	0.110	30.000	29.883	0.065	30.000	29.959	0.020	30.000	29.980	0.007	30.000	29.987	0.000

		LOCATIONAL TRANSN.			LOCATIONAL TRANSN.			LOCATIONAL INTERF.			MEDIUM DRIVE			FORCE		
BASIC SIZE		Hole H7	Shaft k6	Fit	Hole H7	Shaft n6	Fit	Hole H7	Shaft p6	Fit	Hole H7	Shaft s6	Fit	Hole H7	Shaft u6	Fit
1	MAX	1.010	1.006	0.010	1.010	1.010	0.006	1.010	1.012	0.004	1.010	1.020	-0.004	1.010	1.024	-0.008
	MIN	1.000	1.000	-0.006	1.000	1.004	-0.010	1.000	1.006	-0.012	1.000	1.014	-0.020	1.000	1.018	-0.024
1.2	MAX	1.210	1.206	0.010	1.210	1.210	0.006	1.210	1.212	0.004	1.210	1.220	-0.004	1.210	1.224	-0.008
	MIN	1.200	1.200	-0.006	1.200	1.204	-0.010	1.200	1.206	-0.012	1.200	1.214	-0.020	1.200	1.218	-0.024
1.6	MAX	1.610	1.606	0.010	1.610	1.610	0.006	1.610	1.612	0.004	1.610	1.620	-0.004	1.610	1.624	-0.008
	MIN	1.600	1.600	-0.006	1.600	1.604	-0.010	1.600	1.606	-0.012	1.600	1.614	-0.020	1.600	1.618	-0.024
2	MAX	2.010	2.006	0.010	2.010	2.010	0.006	2.010	2.012	0.004	2.010	2.020	-0.004	2.010	2.024	-0.008
	MIN	2.000	2.000	-0.006	2.000	2.004	-0.010	2.000	2.006	-0.012	2.000	2.014	-0.020	2.000	2.018	-0.024
2.5	MAX	2.510	2.506	0.010	2.510	2.510	0.006	2.510	2.512	0.004	2.510	2.520	-0.004	2.510	2.524	-0.008
	MIN	2.500	2.500	-0.006	2.500	2.504	-0.010	2.500	2.506	-0.012	2.500	2.514	-0.020	2.500	2.518	-0.024
3	MAX	3.010	3.006	0.010	3.010	3.010	0.006	3.010	3.012	0.004	3.010	3.020	-0.004	3.010	3.024	-0.008
	MIN	3.000	3.000	-0.006	3.000	3.004	-0.010	3.000	3.006	-0.012	3.000	3.014	-0.020	3.000	3.018	-0.024
4	MAX	4.012	4.009	0.011	4.012	4.016	0.004	4.012	4.020	0.000	4.012	4.027	-0.007	4.012	4.031	-0.011
	MIN	4.000	4.001	-0.009	4.000	4.008	-0.016	4.000	4.012	-0.020	4.000	4.019	-0.027	4.000	4.023	-0.031
5	MAX	5.012	5.009	0.011	5.012	5.016	0.004	5.012	5.020	0.000	5.012	5.027	-0.007	5.012	5.031	-0.011
	MIN	5.000	5.001	-0.009	5.000	5.008	-0.016	5.000	5.012	-0.020	5.000	5.019	-0.027	5.000	5.023	-0.031
6	MAX	6.012	6.009	0.011	6.012	6.016	0.004	6.012	6.020	0.000	6.012	6.027	-0.007	6.012	6.031	-0.011
	MIN	6.000	6.001	-0.009	6.000	6.008	-0.016	6.000	6.012	-0.020	6.000	6.019	-0.027	6.000	6.023	-0.031
8	MAX	8.015	8.010	0.014	8.015	8.019	0.005	8.015	8.024	0.000	8.015	8.032	-0.008	8.015	8.037	-0.013
	MIN	8.000	8.001	-0.010	8.000	8.010	-0.019	8.000	8.015	-0.024	8.000	8.023	-0.032	8.000	8.028	-0.037
10	MAX	10.015	10.010	0.014	10.015	10.019	0.005	10.015	10.024	0.000	10.015	10.032	-0.008	10.015	10.037	-0.013
	MIN	10.000	10.001	-0.010	10.000	10.010	-0.019	10.000	10.015	-0.024	10.000	10.023	-0.032	10.000	10.028	-0.037
12	MAX	12.018	12.012	0.017	12.018	12.023	0.006	12.018	12.029	0.000	12.018	12.039	-0.010	12.018	12.044	-0.015
	MIN	12.000	12.001	-0.012	12.000	12.012	-0.023	12.000	12.018	-0.029	12.000	12.028	-0.039	12.000	12.033	-0.044
16	MAX	16.018	16.012	0.017	16.018	16.023	0.006	16.018	16.029	0.000	16.018	16.039	-0.010	16.018	16.044	-0.015
	MIN	16.000	16.001	-0.012	16.000	16.012	-0.023	16.000	16.018	-0.029	16.000	16.028	-0.039	16.000	16.033	-0.044
20	MAX	20.021	20.015	0.019	20.021	20.028	0.006	20.021	20.035	-0.001	20.021	20.048	-0.014	20.021	20.054	-0.020
	MIN	20.000	20.002	-0.015	20.000	20.015	-0.028	20.000	20.022	-0.035	20.000	20.035	-0.048	20.000	20.041	-0.054
25	MAX	25.021	25.015	0.019	25.021	25.028	0.006	25.021	25.035	-0.001	25.021	25.048	-0.014	25.021	25.061	-0.027
	MIN	25.000	25.002	-0.015	25.000	25.015	-0.028	25.000	25.022	-0.035	25.000	25.035	-0.048	25.000	25.048	-0.061
30	MAX	30.021	30.015	0.019	30.021	30.028	0.006	30.021	30.035	-0.001	30.021	30.048	-0.014	30.021	30.061	-0.027
	MIN	30.000	30.002	-0.015	30.000	30.015	-0.028	30.000	30.022	-0.035	30.000	30.035	-0.048	30.000	30.048	-0.061

TABLE 8-5
Preferred Shaft Basis Fits for Diameters to 30 mm

BASIC SIZE		LOOSE RUNNING			FREE RUNNING			CLOSE RUNNING			SLIDING			LOCATIONAL CLEARANCE		
		Hole C11	Shaft h11	Fit	Hole D9	Shaft h9	Fit	Hole F8	Shaft h7	Fit	Hole G7	Shaft h6	Fit	Hole H7	Shaft h6	Fit
1	MAX	1.120	1.000	0.180	1.045	1.000	0.070	1.020	1.000	0.030	1.012	1.000	0.018	1.010	1.000	0.016
	MIN	1.060	0.940	0.060	1.020	0.975	0.020	1.006	0.990	0.006	1.002	0.994	0.002	1.000	0.994	0.000
1.2	MAX	1.320	1.200	0.180	1.245	1.200	0.070	1.220	1.200	0.030	1.212	1.200	0.018	1.210	1.200	0.016
	MIN	1.260	1.140	0.060	1.220	1.175	0.020	1.206	1.190	0.006	1.202	1.194	0.002	1.200	1.194	0.000
1.6	MAX	1.720	1.600	0.180	1.645	1.600	0.070	1.620	1.600	0.030	1.612	1.600	0.018	1.610	1.600	0.016
	MIN	1.660	1.540	0.060	1.620	1.575	0.020	1.606	1.590	0.006	1.602	1.594	0.002	1.600	1.594	0.000
2	MAX	2.120	2.000	0.180	2.045	2.000	0.070	2.020	2.000	0.030	2.012	2.000	0.018	2.010	2.000	0.016
	MIN	2.060	1.940	0.060	2.020	1.975	0.020	2.006	1.990	0.006	2.002	1.994	0.002	2.000	1.994	0.000
2.5	MAX	2.620	2.500	0.180	2.545	2.500	0.070	2.520	2.500	0.030	2.512	2.500	0.018	2.510	2.500	0.016
	MIN	2.560	2.440	0.060	2.520	2.475	0.020	2.506	2.490	0.006	2.502	2.494	0.002	2.500	2.494	0.000
3	MAX	3.120	3.000	0.180	3.045	3.000	0.070	3.020	3.000	0.030	3.012	3.000	0.018	3.010	3.000	0.016
	MIN	3.060	2.940	0.060	3.020	2.975	0.020	3.006	2.990	0.006	3.002	2.994	0.002	3.000	2.994	0.000
4	MAX	4.145	4.000	0.220	4.060	4.000	0.090	4.028	4.000	0.040	4.016	4.000	0.024	4.012	4.000	0.020
	MIN	4.070	3.925	0.070	4.030	3.970	0.030	4.010	3.988	0.010	4.004	3.992	0.004	4.000	3.992	0.000
5	MAX	5.145	5.000	0.220	5.060	5.000	0.090	5.028	5.000	0.040	5.016	5.000	0.024	5.012	5.000	0.020
	MIN	5.070	4.925	0.070	5.030	4.970	0.030	5.010	4.988	0.010	5.004	4.992	0.004	5.000	4.992	0.000
6.	MAX	6.145	6.000	0.220	6.060	6.000	0.090	6.028	6.000	0.040	6.016	6.000	0.024	6.012	6.000	0.020
	MIN	6.070	5.925	0.070	6.030	5.970	0.030	6.010	5.988	0.010	6.004	5.992	0.004	6.000	5.992	0.000
8	MAX	8.170	8.000	0.260	8.076	8.000	0.112	8.035	8.000	0.050	8.020	8.000	0.029	8.015	8.000	0.024
	MIN	8.080	7.910	0.080	8.040	7.964	0.040	8.013	7.985	0.013	8.005	7.991	0.005	8.000	7.991	0.000
10	MAX	10.170	10.000	0.260	10.076	10.000	0.112	10.035	10.000	0.050	10.020	10.000	0.029	10.015	10.000	0.024
	MIN	10.080	9.910	0.080	10.040	9.964	0.040	10.013	9.985	0.013	10.005	9.991	0.005	10.000	9.991	0.000
12	MAX	12.205	12.000	0.315	12.093	12.000	0.136	12.043	12.000	0.061	12.024	12.000	0.035	12.018	12.000	0.029
	MIN	12.095	11.890	0.095	12.050	11.957	0.050	12.016	11.982	0.016	12.006	11.989	0.006	12.000	11.989	0.000
16	MAX	16.205	16.000	0.315	16.093	16.000	0.136	16.043	16.000	0.061	16.024	16.000	0.035	16.018	16.000	0.029
	MIN	16.095	15.890	0.095	16.050	15.957	0.050	16.016	15.982	0.016	16.006	15.989	0.006	16.000	15.989	0.000
20	MAX	20.240	20.000	0.370	20.117	20.000	0.169	20.053	20.000	0.074	20.028	20.000	0.041	20.021	20.000	0.034
	MIN	20.110	19.870	0.110	20.065	19.948	0.065	20.020	19.979	0.020	20.007	19.987	0.007	20.000	19.987	0.000
25	MAX	25.240	25.000	0.370	25.117	25.000	0.169	25.053	25.000	0.074	25.028	25.000	0.041	25.021	25.000	0.034
	MIN	25.110	24.870	0.110	25.065	24.948	0.065	25.020	24.979	0.020	25.007	24.987	0.007	25.000	24.987	0.000
30	MAX	30.240	30.000	0.370	30.117	30.000	0.169	30.053	30.000	0.074	30.028	30.000	0.041	30.021	30.000	0.034
	MIN	30.110	29.870	0.110	30.065	29.948	0.065	30.020	29.979	0.020	30.007	29.987	0.007	30.000	29.987	0.000

BASIC SIZE		LOCATIONAL TRANSN.			LOCATIONAL TRANSN.			LOCATIONAL INTERF.			MEDIUM DRIVE			FORCE		
		Hole K7	Shaft h6	Fit	Hole N7	Shaft h6	Fit	Hole P7	Shaft h6	Fit	Hole S7	Shaft h6	Fit	Hole U7	Shaft h6	Fit
1	MAX	1.000	1.000	0.006	0.996	1.000	0.002	0.994	1.000	0.000	0.986	1.000	−0.008	0.982	1.000	−0.012
	MIN	0.990	0.994	−0.010	0.986	0.994	−0.014	0.984	0.994	−0.016	0.976	0.994	−0.024	0.972	0.994	−0.028
1.2	MAX	1.200	1.200	0.006	1.196	1.200	0.002	1.194	1.200	0.000	1.186	1.200	−0.008	1.182	1.200	−0.012
	MIN	1.190	1.194	−0.010	1.186	1.194	−0.014	1.184	1.194	−0.016	1.176	1.194	−0.024	1.172	1.194	−0.028
1.6	MAX	1.600	1.600	0.006	1.596	1.600	0.002	1.594	1.600	0.000	1.586	1.600	−0.008	1.582	1.600	−0.012
	MIN	1.590	1.594	−0.010	1.586	1.594	−0.014	1.584	1.594	−0.016	1.576	1.594	−0.024	1.572	1.594	−0.028
2	MAX	2.000	2.000	0.006	1.996	2.000	0.002	1.994	2.000	0.000	1.986	2.000	−0.008	1.982	2.000	−0.012
	MIN	1.990	1.994	−0.010	1.986	1.994	−0.014	1.984	1.994	−0.016	1.976	1.994	−0.024	1.972	1.994	−0.028
2.5	MAX	2.500	2.500	0.006	2.496	2.500	0.002	2.494	2.500	0.000	2.486	2.500	−0.008	2.482	2.500	−0.012
	MIN	2.490	2.494	−0.010	2.486	2.494	−0.014	2.484	2.494	−0.016	2.476	2.494	−0.024	2.472	2.494	−0.028
3	MAX	3.000	3.000	0.006	2.996	3.000	0.002	2.994	3.000	0.000	2.986	3.000	−0.008	2.982	3.000	−0.012
	MIN	2.990	2.994	−0.010	2.986	2.994	−0.014	2.984	2.994	−0.016	2.976	2.994	−0.024	2.972	2.994	−0.028
4	MAX	4.003	4.000	0.011	3.996	4.000	0.004	3.992	4.000	0.000	3.985	4.000	−0.007	3.981	4.000	−0.011
	MIN	3.991	3.992	−0.009	3.984	3.992	−0.016	3.980	3.992	−0.020	3.973	3.992	−0.027	3.969	3.992	−0.031
5	MAX	5.003	5.000	0.011	4.996	5.000	0.004	4.992	5.000	0.000	4.985	5.000	−0.007	4.981	5.000	−0.011
	MIN	4.991	4.992	−0.009	4.984	4.992	−0.016	4.980	4.992	−0.020	4.973	4.992	−0.027	4.969	4.992	−0.031
6	MAX	6.003	6.000	0.011	5.996	6.000	0.004	5.992	6.000	0.000	5.985	6.000	−0.007	5.981	6.000	−0.011
	MIN	5.991	5.992	−0.009	5.984	5.992	−0.016	5.980	5.992	−0.020	5.973	5.992	−0.027	5.969	5.992	−0.031
8	MAX	8.005	8.000	0.014	7.996	8.000	0.005	7.991	8.000	0.000	7.983	8.000	−0.008	7.978	8.000	−0.013
	MIN	7.990	7.991	−0.010	7.981	7.991	−0.019	7.976	7.991	−0.024	7.968	7.991	−0.032	7.963	7.991	−0.037
10	MAX	10.005	10.000	0.014	9.996	10.000	0.005	9.991	10.000	0.000	9.983	10.000	−0.008	9.978	10.000	−0.013
	MIN	9.990	9.991	−0.010	9.981	9.991	−0.019	9.976	9.991	−0.024	9.968	9.991	−0.032	9.963	9.991	−0.037
12	MAX	12.006	12.000	0.017	11.995	12.000	0.006	11.989	12.000	0.000	11.979	12.000	−0.010	11.974	12.000	−0.015
	MIN	11.988	11.989	−0.012	11.977	11.989	−0.023	11.971	11.989	−0.029	11.961	11.989	−0.039	11.956	11.989	−0.044
16	MAX	16.006	16.000	0.017	15.995	16.000	0.006	15.989	16.000	0.000	15.979	16.000	−0.010	15.974	16.000	−0.015
	MIN	15.988	15.989	−0.012	15.977	15.989	−0.023	15.971	15.989	−0.029	15.961	15.989	−0.039	15.956	15.989	−0.044
20	MAX	20.006	20.000	0.019	19.993	20.000	0.006	19.986	20.000	−0.001	19.973	20.000	−0.014	19.967	20.000	−0.020
	MIN	19.985	19.987	−0.015	19.972	19.987	−0.028	19.965	19.987	−0.035	19.952	19.987	−0.048	19.946	19.987	−0.054
25	MAX	25.006	25.000	0.019	24.993	25.000	0.006	24.986	25.000	−0.001	24.973	25.000	−0.014	24.960	25.000	−0.027
	MIN	24.985	24.987	−0.015	24.972	24.987	−0.028	24.965	24.987	−0.035	24.952	24.987	−0.048	24.939	24.987	−0.061
30	MAX	30.006	30.000	0.019	29.993	30.000	0.006	29.986	30.000	−0.001	29.973	30.000	−0.014	29.960	30.000	−0.027
	MIN	29.985	29.987	−0.015	29.972	29.987	−0.028	29.965	29.987	−0.035	29.952	29.987	−0.048	29.939	29.987	−0.061

in various categories based on their size. Small shafts are then matched with small holes, medium shafts with medium holes, and so forth. In this manner, satisfactory fits that function properly can be obtained.

SURFACE TEXTURE

The proper functioning of mating parts as well as their service life are affected by the texture of the surface. **Surface texture** refers to repetitive or random deviations from the true surface. These deviations are described by roughness, waviness, lay, and flaws. They are illustrated in Figure 8-38.

Roughness is the relatively finely spaced surface irregularities that are produced by the manufacturing process. **Roughness height** is the average deviation from the mean plane of the surface. It is expressed in micrometers (microinches). **Roughness width** is the distance between successive peaks and is measured in millimeters (inches). **Roughness width cutoff** is the largest spacing of repetitive irregularities to be included in the measurement of the average roughness height. If it is not specified, a value of 0.8 mm (0.030 inch) is assumed.

Waviness is the widely spaced surface undulations that exceed the roughness irregularities as shown in Figure 8-38. Roughness may be viewed as superimposed on the wavy surface. **Waviness height** is the peak-to-valley distance between waves and is specified in millimeters (inches). **Waviness width** is the spacing between wave peaks or wave valleys and is measured in millimeters

Figure 8-38
Characteristics of surface texture.

(inches). **Flaws** are irregularities such as cracks, checks, blowholes, scratches, and so forth, that occur at one place or at relatively infrequent or widely varying intervals on the surface. **Lay** is the direction of the surface pattern caused by the manufacturing process. Figure 8-39 shows six different lay patterns. Also shown is the appropriate symbol.

A surface whose finish is to be specified is marked with the symbol as shown in Figure 8-40. The point of the "V" is on the line representing the surface, on the extension line, or on a leader that points to the surface. It is not necessary to use the entire symbol or specify all surface texture characteristics. The minimum basic symbol is shown at (b). It is constructed with a 60° included angle and the right leg twice as long as the left. When the removal of material by machining is necessary, a horizontal bar is added at the top of the short leg as shown at (c). At (d), the added value indicates the amount of stock in millimeters (inches) to be removed. When material removal is prohibited, a small circle is added in the "V" as shown at (e). This circle indicates that the surface is to be produced without removal of material by some production process such as die casting, powder metallurgy, injection molding, forging, or extruded shape.

Figure 8-41 shows the expected roughness height for some common manufacturing processes. Roughness height is sometimes specified in the symbol by two numbers—one on top of the other. In those cases, the values represent the maximum and minimum average roughness height.

Parallel to boundry   Perpendicular to boundry

Both directions   Multidirectional

Circular   Radial

Figure 8-39
Lay patterns and symbols.

(a)

Figure 8-40
Surface texture symbols.

ROUGHNESS AVERAGE — MICROMETERS μm (MICROINCHES μ in)

PROCESS	50 (2000)	25 (1000)	12.5 (500)	6.3 (250)	3.2 (125)	1.6 (63)	0.80 (32)	0.40 (16)	0.20 (8)	0.10 (4)	0.05 (2)	0.025 (1)	0.012 (0.5)
FLAME CUTTING													
SNAGGING													
SAWING													
PLANING, SHAPING													
DRILLING													
CHEMICAL MILLING													
ELECT. DISCHARGE MACH													
MILLING													
BROACHING													
REAMING													
ELECTRON BEAM													
LASER													
ELECTRO-CHEMICAL													
BORING, TURNING													
BARREL FINISHING													
ELECTROLYTIC GRINDING													
ROLLER BURNISHING													
GRINDING													
HONING													
ELECTRO-POLISH													
POLISHING													
LAPPING													
SUPERFINISHING													
SAND CASTING													
HOT ROLLING													
FORGING													
PERM MOLD CASTING													
INVESTMENT CASTING													
EXTRUDING													
COLD ROLLING, DRAWING													
DIE CASTING													

The ranges shown above are typical of the processes listed.

Higher or lower values may be obtained under special conditions.

KEY � Average Application

▨ Less Frequent Application

Figure 8-41
Roughness height of common processes.

SUMMARY

In summary, one must know or do the following:

1. Tolerance is the total amount a specific dimension is permitted to vary.

2. Limit dimensioning expresses the tolerance by giving an upper value and a lower value.

3. Plus and minus dimensioning expresses the tolerance as a deviation from a basic value. The tolerance is unilateral or bilateral.

4. Tolerance accumulation is minimized by the direct dimensioning technique and maximized by chain dimensioning.

5. Different manufacturing processes are capable of different degrees of accuracy.

6. Geometric tolerancing is described as a means of specifying the geometry of a part feature.

7. Specifying the geometric tolerance is simplified by the use of geometric characteristic symbols, datums, basic modifiers, and feature control frames.

8. A basic dimension is a theoretical exact dimension without tolerance.

9. There are three material condition modifiers: MMC, LMC, and RFS.

10. Rule 1 states that where only a feature of size is specified, the limits of size prescribe the amount of variation permitted in its form.

11. Rule 2 states that where a tolerance of position is specified, then RFS, LMC, or MMC must be specified with respect to the tolerance, the datum, or both.

12. Rule 3 states that for other than tolerance of position RFS applies with respect to the tolerance, the datum, or both, unless MMC is specified.

13. Datum planes are associated with the process or inspection equipment.

14. The datum reference frame is three mutually perpendicular planes.

15. The three planes are called the primary, secondary, and tertiary.

16. A primary datum plane requires three points of contact; a secondary datum plane requires two points of contact; and a tertiary datum plane requires one point of contact.

17. Datum targets are designated points, lines, or areas of contact. They are usually locating pins.

18. Flatness is the condition of a surface having all elements in one plane.

19. Straightness is the condition in which an element of a surface or an axis is a straight line.

20. Circularity is a condition of a surface of revolution where all points of the surface intersected by any plane perpendicular to a common axis or passing through a common center are equidistant from the center.

21. Cylindricity is a condition of a surface of revolution in which all points of the surface are equidistant from a common axis.

22. Profile tolerancing is a method of controlling irregular surfaces, lines, arcs, or unusual shapes as well as regular shapes.

23. Angularity is the condition of a surface or axis at a specified angle from a datum plane or axis.

24. Perpendicularity is the condition of a surface, center plane, or axis at a right angle to a datum plane or axis.

25. Parallelism is the condition of a surface equidistant at all points from a datum plane or an axis equidistant along its length to a datum axis.

26. Positional tolerance defines a zone in which the center, axis, or center plane is permitted to vary from the exact location, which is referred to as its true position.

27. Where a hole is at MMC, its axis must fall within a cylindrical tolerance zone whose axis is located at the true position.

28. Concentricity is the condition where the axis of all cross-sectional elements of a surface of revolution are common to the axis of a datum reference.

29. Symmetry is a condition in which a feature or

features are symmetrically disposed about the center plane of a datum feature.

30. Circular runout provides control of circular elements of a surface.

31. Total runout provides combination control of all surface elements.

32. Whenever two parts are designed to function together, they are mating parts.

33. The specification of a mating part requires the basic size, fundamental deviation, and international tolerance grade.

34. There are three types of fits: clearance, interference, and transition.

35. The hole basis system is the most common system, because standard drill bits, reamers, and broaches are used, whereas shafts can be machined to any size.

36. Surface texture refers to repetitive or random deviations from the true surface.

37. Surface texture is described by roughness, waviness, lay, and flaws.

FASTENING

TECHNIQUES

Upon completion of this chapter, the student is expected to:

- Identify the types of thread representation found on a drawing.

- Identify the different series of screw threads.

- Identify three thread classes.

- Identify an external and internal thread type found on a drawing.

- Be able to read the specification of a screw thread as it is shown on a drawing.

- Identify the basic weld symbols found on a drawing.

- Identify four types of welds.

- Identify the six fundamental welds.

- Identify the five types of weld joints.

- Identify at least three techniques of fastening that may be considered nontraditional.

chapter

9

INTRODUCTION

Individual piece parts and subassemblies that make up an end product are joined together by different fasteners. The type of fastener depends on the function of the device and whether the fastener is removable or permanent. Typical removable fasteners are bolts and nuts, screws, pins, and keys; typical permanent fasteners are rivets, nails, staples, drive pins, adhesives, and welding. The intent of this chapter is to illustrate how the different fastening techniques are specified on a drawing. Standards have been developed for some of the fasteners, while for others a local or general note is required.

THREADED FASTENERS

Screw threads are features vital to our industrial life. They are designed for many applications. Three basic applications of screw threads are: (1) holding parts together, (2) adjusting parts with reference to each other, and (3) transmitting power.

A thread is a uniform helical groove cut on or in a cylinder or cone. Thread cutting on a lathe is one of the most exacting lathe operations. A single-point cutting tool is used in lathe operations; however, a threading die for external thread cutting and a standard tap and die set for internal thread cutting can be used. Threads may also be ground or rolled, using a special machine. There are many different types and forms of threaded fasteners. A knowledge of thread terms, thread form thread representation, thread series, thread classes, and specification of screw threads is needed. Threaded fasteners are manufactured in standard types and sizes. These standard parts may be purchased at a nominal cost; specially made sizes and forms are much more expensive.

Threads may be either internal or external. Figure 9-1 shows an example of an external and an internal thread as they might appear on an orthographic drawing. External threads are those on a round shaft, and internal threads are those inside a round hole. The screw thread is in the form of a spiral around the shaft or in the hole, and these spirals must match if the two threads are to fit together.

(a) External threads

(b) Internal threads

Figure 9-1
Screw threads.

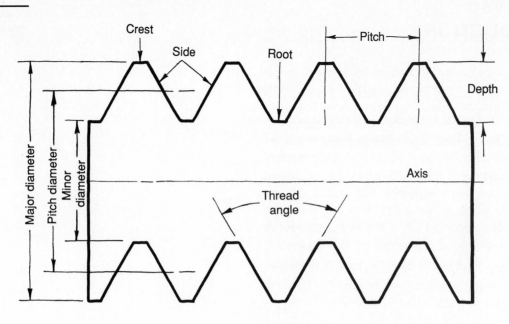

Figure 9-2
Nomenclature of screw threads.

Terminology

Following are some of the common terms and definitions for screw threads as shown in Figure 9-2.

External or male thread. A thread on the outside of a cylinder or cone.

Internal or female thread. A thread on the inside of a hollow cylinder or bore.

Pitch. The distance from a given point on one thread to a similar point on a thread next to it, measured parallel to the axis of the cylinder. The pitch is equal to 1 divided by the number of threads per millimeter or number of threads per inch.

Lead. The distance a screw thread advances axially in one complete revolution. On a single-thread screw, the lead is equal to the pitch. On a double-thread screw, the lead is equal to twice the pitch, and on a triple-thread screw, the lead is equal to three times the pitch (see Figure 9-6).

Crest (also called "flat"). The top or outer surface of the thread joining the two sides.

Root. The bottom or inner surface joining the sides of two adjacent threads.

Side. The side of a thread is the surface that connects the crest and the root.

Thread angle. The angle between the sides of adjacent threads, measured in an axial plane.

Depth. The depth of a thread is the distance between the crest and root of a thread measured perpendicular to the axis.

Major diameter. The largest diameter of a screw thread.

Minor diameter. The smallest diameter of a screw thread.

Pitch diameter. The diameter of an imaginary cylinder formed where the width of the groove is equal to one-half of the pitch. This is the critical dimension of threading as the fit of the thread is determined by the pitch diameter.

Forms

The type of thread form specified by the design engineer is dependent on its application. Thread forms used in U.S. industry are based on approved U.S. standards. Figure 9-3 shows the profile of several commonly used threads. The unified form has been the standard in the English system; however, it is gradually being replaced by the metric form, which

**Figure 9-3
Thread forms.**

has slightly less thread depth. Sharp V threads are used to some extent where adjustment and holding power are essential. The knuckle thread is commonly found on light bulbs and bottle caps and is usually rolled from sheet metal. Acme and B&S worm threads are used for transmitting power.

Left and Right Hand

A thread may be either right hand or left hand, depending on the application. Since most threads in use today are right hand, there is no indication to this effect on the drawing. If a thread is left hand, the letters LH are placed at the end of the thread note. Figure 9-4 shows a right- and left-hand thread as they would appear in detail on a drawing. Figure 9-5 shows the application of right-hand and left-

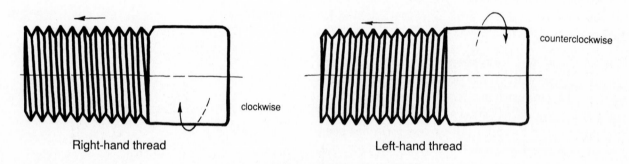

Right-hand thread   Left-hand thread

Figure 9-4
Right- and left-hand threads.

Grinding force tends to reverse the rotation of the wheel. The grinding wheel must act to tighten the nut.

Left-hand thread   Right-hand thread

Figure 9-5
Application of right- and
left-hand threads.

hand threads. The thread must be such that tool pressure exerted on the grinding wheel will tend to tighten the nut holding the wheel on the shaft. If the thread is the wrong hand, the nut will loosen and the wheel will fall off. The same theory applies to the rear-axle nuts on an automobile, where one is right-handed and the other is left-handed. It also applies to bicycle pedals.

Lead

Lead is the distance a screw thread advances in one turn, measured parallel to the axis. Figure 9-6 shows the relationship of pitch to lead for single and multiple threads. In a single thread, the lead is equal to the pitch; in a double thread, the lead is equal to twice the pitch; and in a triple thread, the lead is equal to three times the pitch.

Figure 9-7 shows an application of multiple threads. In single threads the lead is equal to the pitch, and in one turn of the screw, the screw advances one pitch. The pitch (P) is equal to 1 divided by the number of threads per inch. In the single-thread application, we have 8 threads per inch; therefore, with a complete turn of the handle, the clamping head will advance $\frac{1}{8}$ inch. For the triple thread, a complete turn of the handle will advance the clamping head $\frac{3}{8}$ inch. Single threads provide finer adjustment and better power transmission, while multiple threads provide rapid movement. Whenever double leads or triple leads are used, quick motion is obtained but not great power.

Figure 9-6
Multiple screw threads.

Single thread

Advantages:
Finer adjustment
Better power transmission

Disadvantage:
Slow moving

Triple thread

Advantage:
Rapid movement

Disadvantages:
Coarse adjustment
Poorer power transmission

Figure 9-7
Application of multiple threads.

Detail

Schematic

Simplified

External   Internal

Figure 9-8
Screw thread representation.

Representation

Threads may be represented on drawings using one of three conventional methods: detailed, schematic, and simplified. To save drafting time, the schematic or simplified conventions are used. The detailed convention is sometimes used to show the geometry of a thread form as a portion of a greatly enlarged detail on a drawing. Figure 9-8 illustrates the representation for external threads and for internal threads.

Unified Series (English)

There are 11 standard series of threads listed under the unified national form. There are three series with graded pitches and eight series with constant pitches. Table 9-1 shows the series for basic major diameters between 0.0600 and 1.5000 inches.

The unified national coarse (UNC) thread series is a general-purpose thread for screws, bolts, and nuts where a finer thread is not required. The unified national fine (UNF) thread series has a greater number of threads per inch and is used where the length of the threaded engagement is short and a small lead angle is desired. This type of thread finds application in the aircraft and automotive industries. The unified national extra fine series (UNEF) has application where even shorter lengths

TABLE 9-1
Unified Standard Screw Thread Series

| SIZES | | BASIC MAJOR DIAMETER | THREADS PER INCH | | | | | | | | | | | SIZE |
| | | | SERIES WITH GRADED PITCHES | | | SERIES WITH CONSTANT PITCHES | | | | | | | | |
PRIMARY	SECONDARY		COARSE UNC	FINE UNF	EXTRA FINE UNEF	4UN	6UN	8UN	12UN	16UN	20UN	28UN	32UN	
0		0.0600	—	80	—	—	—	—	—	—	—	—	—	0
	1	0.0730	64	72	—	—	—	—	—	—	—	—	—	1
2		0.0860	56	64	—	—	—	—	—	—	—	—	—	2
	3	0.0990	48	56	—	—	—	—	—	—	—	—	—	3
4		0.1120	40	48	—	—	—	—	—	—	—	—	—	4
5		0.1250	40	44	—	—	—	—	—	—	—	—	—	5
6		0.1380	32	40	—	—	—	—	—	—	—	—	UNC	6
8		0.1640	32	36	—	—	—	—	—	—	—	—	UNC	8
10		0.1900	24	32	—	—	—	—	—	—	—	—	UNF	10
	12	0.2160	24	28	32	—	—	—	—	—	—	UNF	UNEF	12
1/4		0.2500	20	28	32	—	—	—	—	—	UNC	UNF	UNEF	1/4
5/16		0.3125	18	24	32	—	—	—	—	—	20	28	UNEF	5/16
3/8		0.3750	16	24	32	—	—	—	—	UNC	20	28	UNEF	3/8
7/16		0.4375	14	20	28	—	—	—	—	16	UNF	UNEF	32	7/16
1/2		0.5000	13	20	28	—	—	—	—	16	UNF	UNEF	32	1/2
9/16		0.5625	12	18	24	—	—	—	UNC	16	20	28	32	9/16
5/8		0.6250	11	18	24	—	—	—	12	16	20	28	32	5/8
	11/16	0.6875	—	—	24	—	—	—	12	16	20	28	32	11/16
3/4		0.7500	10	16	20	—	—	—	12	UNF	UNEF	28	32	3/4
	13/16	0.8125	—	—	20	—	—	—	12	16	UNEF	28	32	13/16
7/8		0.8750	9	14	20	—	—	—	12	16	UNEF	28	32	7/8
	15/16	0.9375	—	—	20	—	—	—	12	16	UNEF	28	32	15/16
1		1.0000	8	12	20	—	—	UNC	UNF	16	UNEF	28	32	1
	1 1/16	1.0625	—	—	18	—	—	8	12	16	20	28	—	1 1/16
1 1/8		1.1250	7	12	18	—	—	8	UNF	16	20	28	—	1 1/8
	1 3/16	1.1875	—	—	18	—	—	8	12	16	20	28	—	1 3/16
1 1/4		1.2500	7	12	18	—	—	8	UNF	16	20	28	—	1 1/4
	1 5/16	1.3125	—	—	18	—	—	8	12	16	20	28	—	1 5/16
1 3/8		1.3750	6	12	18	—	UNC	8	UNF	16	20	28	—	1 3/8
	1 7/16	1.4375	—	—	18	—	6	8	12	16	20	28	—	1 7/16
1 1/2		1.5000	6	12	18	—	UNC	8	UNF	16	20	28	—	1 1/2

of thread engagement are required, such as for thin-walled tubes, nuts, ferrules, and couplings.

Of the eight series with constant pitch, 8 UN, 12 UN, and 16 UN are the most common. The general application of the 8-pitch series (8 threads per inch) is for bolts for high-pressure pipe flanges, cylinder head studs, and similar fasteners. The 12-pitch thread series (12 threads per inch) is used in boiler work and for thin nuts on shafts and sleeves in machine construction. The 16-pitch thread series (16 threads per inch) is used where a fine thread is necessary, regardless of the diameter.

Threads are further classified by the tolerance and allowance permitted in their manufacture. **Allowance** is the difference between the largest shaft (MMC) and smallest hole (MMC). It is the tightest fit. A class 1 fit is recommended for screw threads where a loose fit between mating parts permits frequent and quick assembly and disassembly of parts. A class 2 fit represents a high-quality type of thread product and is recommended for nuts, bolts, and screws used in normal applications found in mass-production industries. A class 3 fit has a tight fit allowance between mating parts and has closer tolerances than either a class 1 or 2 fit. This type of fit, because of its precision and cost, should only be used in special cases. The number for the class of fit is usually followed by the letter A or B, indicating an external (A) or internal (B) thread callout.

Metric Thread Series

There are 14 standard series of threads, and these are shown in Table 9-2 for diameters between 1.6 and 26 mm. Note that thread diameters are selected based on an order of preference using columns 1, 2, and 3 in that order.

Threads are further classified by tolerance grades for the crest (major diameter for external and minor diameter for internal) and pitch diameter using the numbers 3 through 9, with 6 being nearly equal to the 2A and 2B classes of fit for the unified system. Grades with numbers less than 6 are used for fine fits and short lengths of engagement; grades with numbers greater than 6 are used for coarse fits and long engagements. Letters are used to specify the allowance and position (internal/external). For internal threads, G = small allowance and H = no allowance; for external threads, e = large allowance, g = small allowance, and h = no allowance.

TABLE 9-2

Metric Screw Thread Series

NOMINAL SIZE (MM)	GRADED PITCHES		THREE-DIAMETER PREFERENCE			CONSTANT PITCHES (MM)											
	COARSE	FINE	1	2	3	6	4	3	2	1.5	1.25	1	0.75	0.5	0.35	0.25	0.2
1.6	0.35	—	1.6	—	—	—	—	—	—	—	—	—	—	—	—	—	0.2
1.8	0.35	—	—	1.8	—	—	—	—	—	—	—	—	—	—	—	—	0.2
2	0.4	—	2	—	—	—	—	—	—	—	—	—	—	—	—	0.25	—
2.2	0.45	—	—	2.2	—	—	—	—	—	—	—	—	—	—	—	0.25	—
2.5	0.45	—	2.5	—	—	—	—	—	—	—	—	—	—	—	0.35	—	—
3	0.5	—	3	—	—	—	—	—	—	—	—	—	—	—	0.35	—	—
3.5	0.6	—	—	3.5	—	—	—	—	—	—	—	—	—	—	0.35	—	—
4	0.7	—	4	—	—	—	—	—	—	—	—	—	—	0.5	—	—	—
4.5	0.75	—	—	4.5	—	—	—	—	—	—	—	—	—	0.5	—	—	—
5	0.8	—	5	—	—	—	—	—	—	—	—	—	—	0.5	—	—	—
5.5	—	—	—	—	5.5	—	—	—	—	—	—	—	—	0.5	—	—	—
6	1	—	6	—	—	—	—	—	—	—	—	—	0.75	—	—	—	—
7	1	—	—	—	7	—	—	—	—	—	—	1	0.75	—	—	—	—
8	1.25	1	8	—	—	—	—	—	—	—	—	1	0.75	—	—	—	—
9	1.25	—	—	—	9	—	—	—	—	—	—	1	0.75	—	—	—	—
10	1.5	1.25	10	—	—	—	—	—	—	—	1.25	1	0.75	—	—	—	—
11	1.5	—	—	—	11	—	—	—	—	—	—	1	0.75	—	—	—	—
12	1.75	1.25	12	—	—	—	—	—	—	1.5	1.25	1	—	—	—	—	—
14	2	1.5	14	—	—	—	—	—	—	1.5	1.25	1	—	—	—	—	—
15	—	—	—	—	15	—	—	—	—	1.5	—	1	—	—	—	—	—
16	2	1.5	16	—	—	—	—	—	—	1.5	—	1	—	—	—	—	—
17	—	—	—	—	17	—	—	—	—	1.5	—	1	—	—	—	—	—
18	2.5	1.5	—	18	—	—	—	—	2	1.5	—	1	—	—	—	—	—
20	2.5	1.5	20	—	—	—	—	—	2	1.5	—	1	—	—	—	—	—
22	2.5	1.5	—	22	—	—	—	—	2	1.5	—	1	—	—	—	—	—
24	3	2	24	—	—	—	—	—	2	1.5	—	1	—	—	—	—	—
25	—	—	—	—	25	—	—	—	2	1.5	—	1	—	—	—	—	—
26	—	—	—	—	26	—	—	—	—	1.5	—	1	—	—	—	—	—

Designation

Screw threads are made to specific dimensions to work properly. To interpret thread dimensions correctly, you must know how threads are specified. The unified, metric, acme, and pipe are the most common thread forms; therefore, this discussion will cover these four.

Unified The specifications of a unified thread is shown with a series of numbers and letters. The pattern for a unified series is shown in Figure 9-9 and must be read from left to right. The first number ($\frac{1}{4}$) shows the nominal size (major diameter) of the thread. This number represents the fractional diameter, the decimal diameter, or the number of the screw size. The second number (20) indicates the number of threads per inch. The thread series (UNC) is the next group of letters. In this example, Unified coarse is indicated. The last number and letter (2A) indicates the class of fit, and whether the thread is internal or external. The letters used are A and B. The A means the thread is external and the B indicates an internal thread.

At other times other information is added to the standard designation as shown at (b). The LH means that the thread is left hand. Threads are considered to be right hand unless an LH designation is used. The last entry (2.50) indicates the length of thread.

(a)

(b)

Figure 9-9
Unified thread designation.

Figure 9-10
Metric thread designation.

Figure 9-11
Acme thread designation.

Metric A series of numbers and letters are also used to designate metric threads. Figure 9-10 shows a complete callout for a metric thread. The letter M for metric followed by the diameter in millimeters (10) and the pitch in millimeters (1.5) separated by the "×" sign. Next is the tolerance grade (6) and the tolerance position (g), which states that it is an external thread with small allowance. Finally is the length of engagement symbol: S for short, N for normal, and L for long. Where the crest and pitch diameter tolerance grades are different, both must be specified. In many cases only the diameter and pitch are specified.

Acme An acme thread follows the same basic format as a unified thread. Acme threads are designated by their nominal diameter, threads per inch, and class fit (Figure 9-11). The principal differences are the thread classes and the word *acme*. The thread classes 2, 3, 4, and 5 are commonly used for acme threads, with a class 2 the loosest and a class 5 the tightest. A general-purpose Acme thread is indicated by the uppercase letter G and centralizing acme threads by the letter C. A stub acme thread is designated, for example, $\frac{3}{8}$-12 stub acme without a letter symbol.

Pipe There are three forms of pipe threads used in industry: regular, dryseal, and aeronautical. The regular pipe thread is the standard for the plumbing trade. Dryseal pipe thread is used for automotive, refrigeration, and hydraulic tube and pipe fittings. The standard in the aerospace industry is aeronauti-

Figure 9-12
Pipe thread designation.

cal pipe thread. The American standard pipe threads specifications are listed in sequence on a print, by nominal size, number of threads per inch, and the symbol for the thread series and form (Figure 9-12). NPT (American standard taper pipe thread) specifies a taper pipe thread and NPS specifies a straight pipe thread.

Typical thread notes for an internal and external feature using English units are shown in Figure 9-13 and metric units in Figure 9-14.

Relief

Some drawings may show a note calling for thread relief. The purpose of thread relief is to relieve the material so that the threading die may pass over the

Figure 9-13
Unified thread specifications.

Figure 9-14
Metric thread specifications.

Figure 9-15
Thread relief.

relief and make the threads complete to the end of the piece, thereby allowing the nut to be screwed tightly against the shoulder. One method to state thread relief is by the width and the depth of groove, as shown in Figure 9-15.

Standard Fasteners

Threads are used to make standard fasteners such as bolts, nuts, cap screws, machine screws, and set-screws. In the specification, the thread information is given first followed by the fastener information.

The most common fastener is the bolt and nut, and they are shown in Figure 9-16. Heads may be regular or heavy depending on application and are usually hexagonal rather than square in shape. Finished bolt heads and nuts are machined on the bearing surface to provide a washer face for the bolt head or nut.

Another common fastener is the cap screw and various types are shown in Figure 9-17. They are used to join two parts together. One of the parts is

Figure 9-16
Bolt and nut.

**Figure 9-17
Cap screws.**

**Figure 9-18
Machine screws.**

an oversized hole and the other is threaded and acts as a nut. When the cap screw is tightened, the two parts are securely held together. Different types of cap screws provide for above- or below-surface mounting.

Machine screws are similar to cap screws and are shown in Figure 9-18. They are particularly adapted to screwing into thin materials, and the smaller screws are threaded to the head. Machine screw nuts are used mainly with the roundhead and flathead types and are hexagonal in shape.

Setscrews are used principally to prevent motion between two parts, such as a pulley and a shaft. The setscrew is screwed through one part, and the point presses on the other. As shown in Figure 9-19, there are six different points and three different head styles.

RIVETS

One example of a permanent fastener is the rivet. Figure 9-20 illustrates the use of permanent fasteners in the form of rivets. The size of the rivet is designated by body diameter and usage varies from small toys to bridges. Many different head styles are avail-

Figure 9-19
Setscrews.

able to suit the application. In structural application, heading is usually done with a die held in a powered hammer and the rivet held up with an anvil or backup bar. Other applications may use a rivet machine, in which the two pieces to be joined have prepierced holes and are located over a mandrel. A rivet is then fed into position down a track and picks up the mandrel; the hammer drives the rivet through the two pieces, striking an anvil on the underside and forming a head, thus securing the two pieces together.

WELDING

Another example of a permanent type of fastener is welding. Welding processes are classified into three main types: (1) arc welding, (2) gas welding, and (3) resistance welding. These main types can be further broken down into subtypes. For example, in arc welding we find such types as metal arc, TIG (tungsten electrode used with an inert gas), MIG (metal inert gas welding), and others. The most common type of gas welding is oxygen and acetylene, which uses a high-temperature flame. Under the resistance welding classification are spot, seam, projection, and high-frequency welding, to name a few.

Two other types of metal joining that are closely allied to welding are brazing and soldering.

Figure 9-20
Rivets.

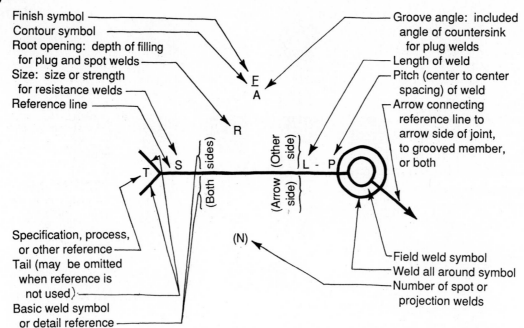

Figure 9-21
Basic weld symbols.

Brazing is the process of joining metals with a non-ferrous filler metal that has a melting point below that of the metals being joined. The American Welding Society (AWS) defines the melting point of the filler as above 427°C (800°F). Below this temperature are the solders. In soldering, two or more pieces of metal are joined by means of a fusible alloy or metal, called solder, which is applied in the molten state.

The AWS has developed and adopted standard procedures for using symbols to indicate location, size, strength, geometry, and other information necessary to describe the weld required. In Figure 9-21 is a welding symbol with the standard placement of information concerning the weld. A full understanding of this standard symbol is not expected, since only a few will have an occasion to use it to make sketches or to interpret the sketch or drawing with the welding symbols. It is expected that, upon seeing the basic welding symbol, you will recognize it. Each item of the basic welding symbol has its place and purpose and identifies some operation or part of the welding process. The symbol in its entirety is seldom necessary. It is modified for particular applications.

Figure 9-22 illustrates common welds, such as the fillet weld, lap weld, and butt weld. The specifi-

Figure 9-22
Types of welds.

cation of the type of weld and the amount of weld are very important in controlling the weld joint.

The welding processes that we have discussed thus far are some of the tried-and-true methods that have been used for years. However, technological advances have produced many new welding processes. These may be classified into three areas: fusion, resistance, and solid state. Two specialized processes in the fusion area are electron-beam welding and laser welding. Welding processes that may be classified as solid-state processes are ultrasonic, explosive, diffusion, and friction processes.

OTHER TECHNIQUES

Several other techniques of fastening are as follows:

1. **Keys.** Basically, keys are a type of fastener that is used to prevent rotation of gears, pulleys, and collars on rotating shafts. The key is a piece of metal that fits into a keyseat in the shaft and keyway in the hub. Figure 9-23 shows the cutter for milling the keyway for a square key and a drawing of a key in assembly.

2. **Locking devices.** Figure 9-24 shows several commonly used locking devices. The castellated nut is slotted to receive a cotter pin that passes through the shaft and slots in the nut. In the center of the figure a jam nut is used. The regular nut is tightened; then the jam nut is tightened against the regular nut. This cre-

Square key

**Figure 9-23
Square key.**

**Figure 9-24
Locking devices.**

Thread cutting
screw

Tapping screw

Carriage bolt

Figure 9-25
Special fasteners, bolts, and nuts.

ates a tension in the threads, tending to prevent the assembly from working loose. In the figure on the right a spring lock washer is shown. The reaction by the lock washer tends to prevent the nut from turning.

3. **Special types of bolts and screws.** Figure 9-25 shows some special types of bolts, screws, nuts, and fasteners that are used in fastening. One of the most common is the sheet metal or self-tapping screw, which is found in automobiles and appliances.

4. **Staples.** We are all familiar with the stapler used in offices, homes, and stores for the permanent joining of paper. This technique of joining materials is also used in manufacturing. Metal stitching uses the same technique as stapling paper: form the staple legs from a wire roll, drive them through the material to be joined, and clench them against an anvil on the opposite side.

5. **Adhesives.** In its never-ending search to make airplanes lighter, the aircraft industry is using the chemical fastening process. Structural bonding of parts is used extensively in automobiles, appliances, and other products. For example, structural adhesives have been used for years to bond brake linings to the brake shoes. The bond develops strengths of 10,000 pounds in shear compared to 2500 pounds shear stress developed by the rivets formerly used.

SUMMARY

In summary, one must know or do the following:

1. Threaded fasteners, rivets, welding, pins, and adhesive bonding are some of the techniques used to join materials together, either permanently or temporarily.

2. Threaded fasteners are specified by the type of thread, size of thread, pitch, and tolerance.

3. Unified, acme, and metric are the three most common types of thread forms.

4. Threads may be either right hand or left hand, external or internal.

5. Lead is the distance a screw thread advances axially in one complete revolution.

6. Common thread terms are pitch, crest, root, depth, angle, major diameter, minor diameter, and pitch diameter.

7. Threads may be represented on a sketch or drawing in one of three conventional ways: detailed, schematic, or simplified. The preferred method is simplified.

8. Threads are presented on a sketch using a symbol designation.

9. Standard fasteners are bolts, nuts, cap screws, machine screws, and setscrews.

10. The three main welding processes are arc, gas, and resistance. Brazing and soldering are closely allied to welding.

11. A welding symbol with the standard placement of information concerning the weld is found on subassembly or assembly drawings.

12. Types of welds are butt, edge, fillet, plug, lap, and double butt.

13. Other types of welding processes are electron beam, laser, ultrasonic, explosive, diffusion, and friction.

14. Other fastening techniques are keys, locking devices, sheet-metal screws, staples, and adhesives.

15. The fastening technique used is specified on a subassembly or assembly drawing by graphical representation, symbols, local notes, or a general note.

WORKING DRAWINGS

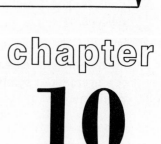

Upon completion of this chapter, the student is expected to:

- Know the different types of drawing.

- Be able to identify the different parts of the title block.

- Be able to identify the different parts of the revision block and know its importance.

- Be able to identify the different parts of the bill of materials and know its use.

- Know the different types of notes and where each is used.

- Be able to interpret a working drawing.

chapter
10

INTRODUCTION

Sketching or drawing is the universal language used by engineers, designers, technologists, technicians, and skilled craftsmen. Whether the drawing is made freehand, with instruments, or by a computer, it is used to convey all necessary information to personnel who will fabricate and assemble the building, aircraft, washing machine, camera, and so on. Copies of drawings are referred to as blueprints, prints, or simply drawings. The color of the copy is a function of the reproduction process.

Reading a drawing involves visualization and interpretation. Visualization is the ability to "see" or envision the size and shape of the object. You may wish to review those chapters (3 through 6) that discussed the different types of views. Operating personnel must be able to interpret symbols, dimensions, notes, and other information.

TYPES OF DRAWINGS

When a machine is manufactured or a structure built, the engineering department provides a set of drawings. These drawings are called working drawings and consist of two basic types: detailed drawings and assembly drawings.

A **detailed drawing** shows a single component or part. In a direct and simple manner, it gives the shape, size, material, finish, tolerance, and so on. The most common types are:

1. Machine drawings

2. Pattern drawings

3. Forging drawings

4. Sheet-metal drawings

5. Electronic drawings

6. Structural drawings

Detailed drawings are not to be confused with detailed views. A detailed view shows a part in the same plane and in the same arrangement, but in greater detail and a larger scale than a principal view.

An **assembly drawing** shows the functional relationships of the various parts of a machine or structure as they fit together to make up an end

Figure 10-1
Working drawing components.

product. The information provided on the assembly drawing is (1) the name of the assembly or sub-assembly, (2) the visual relationship of one part to another to ensure proper assembly of the various parts, and (3) the lists of parts showing number, name, and quantity. Assembly drawings are classified as follows:

1. Design assembly drawings

2. Subassembly drawings

3. Working assembly drawings

4. Checking assembly drawings

5. Installation drawings

6. Outline drawings

7. Illustrative assembly drawings

8. Exploded-view assembly drawings

9. Diagram assembly drawings

10. Plant layout drawings

COMPONENTS OF WORKING DRAWINGS

Working drawings are best analyzed by breaking them down into their components. The components are as follows:

1. Title block

2. Revisions

3. Bill of materials

4. General notes

5. Local notes

6. Size and shape description

Their usual locations on the drawing are illustrated in Figure 10-1.

TITLE BLOCK

The title block is illustrated in Figure 10-2; it is located in the lower right-hand corner of the drawing so that when folded properly the title block may be seen for easy reference and filing. Title blocks provide the basic information about the part or assembly. This information will vary from company to company depending on the specific needs of the company. It includes, but is not limited, to the following information:

1. Name and address of company

2. Name of part, subassembly, or assembly

3. Part number and drawing number, if different

4. Scale (When the equal sign is used, such as 1 = 4, the English system is implied; the colon, 1:4, implies the metric system. In addition, the word METRIC should be prominently given on the drawing.)

5. Paper size or code letter (Table 10-1 gives the standard sizes and code letters for both systems.)

6. Sheet number (sheet 3 of 5)

7. Latest revision number (usually a code letter) and date

8. Date drawn

9. Preparer (designer or drafter)

10. Checker's initials and date

11. Approver's initials and date

12. Miscellaneous information, such as material, heat treatment, finish, estimated weight, and stress analysis

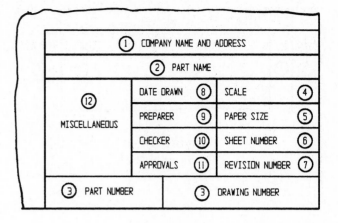

Figure 10-2
Title block.

TABLE 10-1
Metric and U.S. Sheet Sizes

SHEET DESIGNATION	SHEET SIZE (mm)	CORRESPONDING U.S. SHEET SIZE	SHEET SIZE (inches)
A0	841 × 1189	E	34 × 44
A1	594 × 841	D	22 × 34
A2	420 × 594	C	17 × 22
A3	297 × 420	B	11 × 22
A4	210 × 297	A	8.5 × 11

REVISIONS				
ZONE	REV.	DESCRIPTION	DATE	APPROVED
⑤	①	②	③	④

Figure 10-3
Revision block.

REVISION BLOCK

The revision block is usually located in the upper right-hand corner of the drawing. It is shown in Figure 10-3 and is used to record all engineering changes to the original drawing. The revision block provides space for (1) the revision symbol, (2) description of the change, (3) date, and (4) approvals. For large, complicated drawings in which zoning is used to find a particular location, a zone column (5) is used.

Revisions or changes are made for a variety of reasons: improved design, change in manufacturing method, cost reduction, quality, and correction of errors in the original drawing. Whatever the reason, the revision must be documented. These revisions start at the top and proceed downward in a chronological order. A symbol, usually a letter, identifies the change in the revision block, and this same symbol is located adjacent to the change in the drawing.

It is very important that the record of changes be fully documented and the information disseminated to all appropriate personnel. This action will enable operating and inventory control personnel to perform their functions correctly. It is also important to enable service personnel to supply the correct part, knowing the serial number, model number, part number, or other identifying information.

BILL OF MATERIALS

The bill of materials is shown in Figure 10-4. It usually appears immediately above the title block and contains a list of the parts and/or material used on or required by an assembly or installation drawing. The bill of materials is sometimes called parts list, materials list, list of materials, or schedule of parts. Sometimes the bill of materials is listed separately on a computer printout.

Accurate bills of material are extremely important for the production and inventory control system to function effectively. The bill of materials enables the purchasing department to order the quantity of materials needed to produce a given number of assemblies. It enables production control to requisition the required quantities from the storeroom and inventory control to maintain the correct quantity of items.

Figure 10-4
Bill of materials.

Although the information on the bill of materials varies from company to company, it usually contains the (1) part number, (2) description of material specification, (3) quantity required, and (4) the material. Additional information that may be added are industry code numbers and overall dimensions.

NOTES

Notes provide additional information and instructions in an efficient manner, thereby supplementing the graphical representation and the information found in the title block and revisions block. There are two types of notes: general notes and local notes, depending on the application. **General notes** apply to the entire drawing and are usually placed to the left of the title block in a horizontal position. Examples of general notes are:

REMOVE BURRS

ALL FILLETS AND RADII R4

ALL DIMENSIONS ±0.1 (mm)

Local notes or specific notes apply to certain features or areas. They are located near and directed to the feature or area by a leader, as shown in Figure 10-5. Local notes can also be referenced from the field of the drawing by the note number enclosed in an equilateral triangle (sometimes called a *flag*). These are then listed in a convenient location on the drawing.

READING A DRAWING

A drawing cannot be read all at once any more than an entire page of print can be read at a glance. Both must be read a line at a time. Steps in reading a drawing are:

1. Read the title block.

2. Read the revisions.

3. Read the bill of materials (subassembly or assembly)

4. Analyze the part or subassembly or assembly (whichever the case).

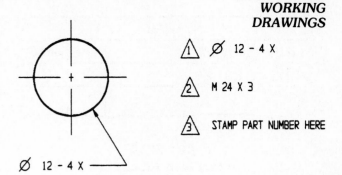

**Figure 10-5
Examples of local notes.**

With this information in mind, let us begin to analyze a working drawing step by step.

TITLE BLOCK EXAMPLE

First we should check the part number and the drawing number, if different. As shown in Figure 10-6 the part number is 648167. The part or assembly name is descriptive, brief, and clearly stated. It starts with the basic name of the part or assembly, is followed by descriptive modifiers, and is read with the descriptive modifiers first. Therefore, the name for this part would be OUTER HARMONIC BALANCER WEIGHT DISK. The drawing scale indicates the size of the drawing as compared with the actual size of the part. The scale of 1 = 1 shows that the drawing is full size. The drawing data found in the title box are January 11, 1980, and that it was prepared by a person with the initials of WBR.

Before a drawing is said to be finished it must be checked for completeness, accuracy, and clarity. Part number 648167 was checked by an individual with the initials of LCN. You can see that two other individuals have given their approval. One of these was probably the drafting supervisor, the other perhaps the design engineer. Normally, before any drawing is used, the above people initial and give their approval to it. Before we fabricate a part, it is necessary that we know the material. To make this part, we find that the material callout is SAE (Society of Automotive Engineers) 1020 CRS (cold-rolled steel). Whether the steel will be sheet stock, strip stock, casting, and so forth, will be determined by the process engineer.

REVISION BLOCK EXAMPLE

Using Figure 10-6 again, we next analyze the revisions. Revision A involves changing the 3.620 dimension to 3.600. The description of this change, which is found under the revision record column, is short and concise. We can also see that revision A was made on May 1, 1981. On July 2, 1982, revision B was made, which affected the .150—.140 dimension of the part. If we look for revision B on the drawing of the part, we find that it is located at the top of the right side view. Notice the B has been

Figure 10-6
Title block and revision example.

circled and appears upper left from the .140—.110 dimension. What this tells us is that the stock thickness was changed from .150—.140 to .140—.110. Revision C changed the dimension from .845—.847 to .895—.897.

BILL OF MATERIALS EXAMPLE

A typical bill of materials is shown in Figure 10-7. First it is noted that one each of part numbers 1, 2, and 3 is required, that is, one base, one yoke, and one barrel. The material is cast iron; C.I. is the abbreviation used. If our plant is a combination gray iron foundry and machining operation, we would be casting parts 1, 2, and 3 and then the necessary machine operations would follow. If, however, we do only machine operations, then the cast-iron parts would be purchased from a foundry, shipped to us, and then machined. Part 4 is a bushing made of

7	1	CAP SC.	1112 ST.	M14 X 2 X 80 HEX
6	2	HEX. NUT	1112 ST.	M10 × 100 JAM
5	2	SET SC.	1112 ST.	M6 X 1 X 12 SLT. CONE PT. SET SC.
4	1	BUSHING	BRONZE	OILITE #1612
3	1	BARREL	C.I.	
2	1	YOKE	C.I.	
1	1	BASE	C.I.	
ITEM	QTY.	NAME	MATERIAL	SPECIFICATIONS
			BILL OF MATERIALS	

Figure 10-7
Bill of materials example.

bronze material with a specification of oilite no. 1612. You will note that only one is required and is purchased. Part number 5 is a setscrew made from 1112 steel with a diameter of 6 mm, 1 thread per millimeter, length of 12 mm, with a slotted head and a cone point. The quantity for part number 5 is 2 per unit assembly; therefore, if 100 unit assemblies were required, we would need 200 of part number 5. Part number 6 is a hexagonal jam nut with a diameter of 10 mm and a length of 100 mm, and part number 7 is a hexagonal head cap screw with a length of 80 mm and a diameter of 14 mm.

The bill of materials must contain sufficient information so that the parts may be manufactured or purchased complete without further information from the people who designed the device or prepared the drawing.

ANALYZE THE PART

In studying an orthographic projection drawing, certain rules of procedure will be helpful in visualizing an object:

1. Scan briefly all views.

2. Study carefully the front view for shape description.

3. Move from the front view to the other views and look for lines that describe the intersections of surfaces, the limits of a surface, or the edge view of a surface.

4. Study one feature at a time in the several views and begin to picture in your mind the shape of the real object.

One cannot read a blueprint by looking at a single view when more than one view is given. Two views will not always describe an object; and when three views are given, all three must be consulted to be sure the shape has been read correctly.

EXAMPLE PROBLEMS

Problem No. 1

Figure 10-8 is an illustration of a two-view detail drawing. The views shown are the front view and the top view. Our analysis should be as follows:

Step 1. Scan briefly all views shown.

Step 2. Study carefully the front view for shape description. The basic shape of this part is a rectangle that is 67 mm wide by 87 mm long. We do not know how thick the object is; therefore, we need to take the next step.

Step 3. Move from the front view to the top view. Here we see that the thickness or depth of this part is 18 mm.

Step 4. Study one feature at a time in the several views and begin to picture in your mind what the object will look like when all operations are complete.

**Figure 10-8
Visualizing the part.**

As we look at the front view, we see that there are four holes of varying sizes. We have a 25 mm diameter hole with 0.02 mm tolerance. The center of this hole is located 28 mm from the right-hand edge of the part and 28 mm from the bottom edge. From all appearances, this hole goes clear through. We can check this by looking at the top view. If we project up from the outer edges of the hole in the front view to the top view, we will see hidden lines. These lines run from one edge (the front) to the other edge (the back) in the top view. Thus we see that the 25 mm diameter hole goes clear through the part. It might be good to note here that the diameter of this hole is very precise. The machinist will drill this hole with a somewhat smaller drill size and then ream to give the final dimension of 25 to 25.02 mm. Note that the hole cannot be any smaller than 25 mm nor larger than 25.02 mm.

The location of the 6 mm hole can be found by looking at the front view and analyzing the dimensions given. We find that the hole is located 28 mm to the left and 20 mm above the large hole. This hole also appears to go clear through the part. We can verify this by projecting up from the front view to the top view. Again we find hidden lines that are as wide as the 6-mm hole, and they too go from front to back. Thus we see that this hole also goes through the part.

Now let us see how you can do on questions relating to the two remaining holes.

- What are the basic sizes of the two holes?

10 mm , 7 mm
- What is the size of the counterbore? How deep is it? *15 mm , 8 mm*

- What is the degree callout for the countersink?

82°

- What diameter will it be machined to?

  12 mm

- What is the distance of the 10-mm hole from the bottom edge?

  50 mm

- Is its vertical centerline located from the right-hand edge or the centerline of the 25-mm-diameter hole? 25 mm HOLE

- What distance is the horizontal centerline of the 10-mm-diameter hole from the horizontal centerline of the 25-mm-diameter hole?

  32 mm

- Is the vertical centerline of the 7-mm-diameter hole 3 mm from the vertical centerline of the 25-mm-diameter hole? YES

- Which direction—right or left?

  LEFT

Figure 10-9
Pattern and machining drawings.

Problem No. 2

Figure 10-9 is a pattern drawing and a machining drawing. The top drawing will give information to the pattern maker as to size and configuration of the part prior to machining. The dimensions as shown on the pattern drawing are the casting size. The extra thick lines represent material that will be removed during the machining.

Our analysis of the pattern drawing should be as follows:

Step 1. Scan briefly all views shown. One view shows a rectangular-appearing object, while the other appears to be a series of concentric circles. Two circles are shown as hidden lines.

Step 2. Study carefully the front view for shape description. Upon analysis, we find a part that is 62 mm wide by 92 mm high.

Step 3. Move from the front view to the other view (top view). When we analyze the top view, we find a series of concentric circles, the largest being 62 mm with a 8 mm core at the center.

Step 4. Study one feature at a time in the views shown and begin to picture in your mind the shape of the real object. When we concern ourselves with the overall appearance of this part, we find that the series of concentric circles will be of varying diameters and lengths. For example, starting at the bottom we have a 38-mm-diameter circle that is 31 mm long. Above this we have a 62 mm diameter circle

185
WORKING
DRAWINGS

(this dimension is found between both views), which is 11 mm long (92-31-10-40). The 56-mm-diameter circle is 10 mm long and the 22-mm-diameter circle is 40 mm long.

When one analyzes a print, it is good practice to have a pencil available (preferably a red one) and check off the dimensions as you analyze them.

Upon further analysis, we see hidden (dashed) lines within the part itself. We know from our previous units of instruction that hidden lines indicate edges, surfaces, and corners that are not visible in a particular view, such as the front view shown in the pattern drawing. The pattern maker, at this point in the analysis, would look at the machining drawing and verify the fact that the part contains a cavity. The machining drawing is a full-sectional view and shows the internal cavity that is to be made in the finished part. There will be times when a machining drawing will not be shown and the pattern drawing will show the sectional view. The internal cavity, as we will find, is of two different sizes. The larger cavity is 29 mm diameter by 42 mm long. At one end, it has a small hole of 8 mm diameter by 5 mm long; at the other end there is a 8-mm-diameter hole. The cavity found in this part will be made by placing a sand core in the mold. The molten metal will surround this core. Once the metal has solidified and the casting is removed from the mold and cleaned, we will find the cavity inside the part. This is the result of the core sand material breaking up and falling through the 8-mm-diameter openings.

Now let us return our attention to the shaded end (front view) of this pattern drawing. Perhaps your eye will focus on the 22-mm dimension. Earlier we had established the fact that this part was made with a series of concentric circles of varying lengths. Therefore, the 22-mm dimension is the diameter of the long projection. How long? Following the extension line at the top of the part out to the right, we find two arrowheads that end at this line. If we follow the arrowhead further from the part and follow it down vertically to its extreme end (at the bottom), we should have seen a figure of 92 mm. This tells us that the overall length of this part is 92 mm. But we want to know the length of the 22-mm-diameter projection. Let us go back up to the top extension line and move to the inner arrow. Following the dimension line down vertically to the next extension line and arrow combination, you should have passed a 40-mm dimension. Your interpreta-

Hydraulic Valve Actuator Assembly
Used in Assembly No. 14-4

Figure 10-10
Subassembly drawing.

tion should tell you that the 22-mm-diameter projection is 40 mm long.

What is the diameter and length of the next surface to be machined? As you look at the pattern drawing, you will note that the diameter is 56 mm by 10 mm long. The machining drawing gives the finished dimensions of the surfaces that are to be machined. These surfaces are identified by a $\checkmark$ on the drawing.

Problem No. 3

Figure 10-10 is a subassembly drawing of a hydraulic valve activator subassembly. As you know, most final assemblies consist of subassemblies and individual piece parts. As you look at this drawing you see that it is made up of three items, 2, 4, and 8. You will also see the relationship between the three as they are to be assembled. In this type of drawing, an outline and description of the parts are given in order that the parts may be assembled in the proper order. A reference is also made as to the next assembly that it will be used on, assembly no. 14-4.

Figure 10-11 is known as a design layout drawing. This type of drawing will show the arrangement of parts, size, and the general overall relationship of the parts, one to another. This type of drawing is used to evaluate the overall design of the product. Once the design has been evaluated, it is then turned over to the detailer, who makes detailed drawings from the layout. As you can see in the drawing, the individual parts that make up the assembly are identified with a number that is not the drawing number. This number would be found on a bill of materials alongside of the part number, part name, and so on. One part that is classed as a standard part, the flathead screw (no. 6), would not be detailed.

Hydraulic Reversing Valve
Assembly No. 14-4

Figure 10-11
Design layout drawing.

SUMMARY

In summary, one must know or do the following:

1. Working drawings consist of two basic types: assembly drawings and detail drawings.

2. Assembly drawings show the parts assembled in the unit or machine and their working relationships.

3. Assembly drawings serve as a place to show the assembled parts and to identify each part.

4. Assembly drawings seldom show all the details of the parts in the assembly.

5. A detail drawing illustrates a single component or part, showing a complete and exact description of its shape, dimensions, and features.

6. The parts of a working drawing are:
   a. Title block
   b. Revision block
   c. Bill of materials (for assembly drawings)
   d. Notes
   e. Detail part, subassembly, assembly

7. The procedure for reading a blueprint is:
   a. Read the title block.
   b. Read the revisions.
   c. Read the bill of materials.
   d. Analyze the part or subassembly or assembly.

CAD

Upon completion of this chapter, the student is expected to:

- **Know the basic components of a computer system.**

- **Know the hardware and software of a CAD system.**

- **Be able to describe the various input methods.**

- **Know the CAD enhancement features.**

- **Know the advantages of CAD.**

INTRODUCTION

This book would not be complete without some mention of the acronym CAD/CAM, which stands for *computer-aided design and computer-aided manufacturing.* To put it simply, CAD/CAM is "the use of computers to aid the design and manufacture process." The conceptualization of a new product, structure, or system, or the improvement thereof, must exist in the mind of the designer before it can become a reality. In most situations, the first step is to communicate ideas by means of a technical sketch.

In the development of an idea, many sketches as well as stress calculations, motion analysis, size of part, material specifications, production methods, and design layouts are made. It is the responsibility of the designer to oversee the preparation of part drawings and part specifications that provide for the many details of production, assembly, and maintenance of the product. To accomplish these tasks, the designer uses freehand sketches to record and communicate ideas rapidly to drafters and support personnel. The ability to convey ideas through sketches and the ability to work with a computer-controlled drawing system is essential.

The thrust of this book has been to familiarize the reader with the basics of technical drawing. Such terms as T-square, 30°-60° triangle compasses, templates, and so forth, were purposely avoided because the authors feel that like the slide rule, drawing instruments will be in limited demand in the years ahead. The ability to sketch, however, will always be needed. It is the age of computer technology and with it, CAD/CAM. The material presented in this chapter is, however, limited to information relating to CAD.

In the original development of CAD/CAM each was treated as a separate discipline and developed independently. Today, CAD/CAM integrates the design and manufacturing function rather than treating them as separate and distinct entities.

Engineering drawings have been and still are an integral part of industry. They are the link between design and manufacturing. Drawings, prepared according to prescribed drafting standards, quickly communicate to manufacturing personnel information for the eventual fabrication of a part or product. It is said, "A picture is worth a thousand words." Actually, a picture is worth much more since the speed of graphic communication can approach a rate of 50,000 times that of reading.

CAD DEFINED

Today it is not necessary to prepare an engineering drawing using conventional tools. Traditionally, drawing instruments have been used to apply india ink on vellum. Now the popular alternative is to prepare the drawing with the aid of a computer. This method is known as *computer-aided design* or *computer-aided design and drawing*, abbreviated CAD or CADD. It has rapidly replaced instrument drawings. Other terms used are *computer-assisted drawing, computer-augmented drawing,* and *computer-automated drawing.* These terms are used synonymously. They will be abbreviated as CAD throughout this chapter.

CAD HISTORY

The CAD/CAM process is an outgrowth of the development of **computer-numerical control** (CNC) machine tools. The start can be found in **numerical control** (NC), where an NC parts program was found to be a nucleus for a manufacturing data base. Users of computer-graphics system in the aircraft industry were generating shapes of complex geometries, with data being created in computer form. From this beginning, the obvious step was to utilize these data to generate manufacturing information. The coupling of these two activities became CAD/CAM—the marriage of computer-aided design and computer-aided manufacturing. Whenever computers are widely used for manufacture, the term CAM is applied to all procedures, where computers assist in the planning and production processes, in inventory control, and for the programming of NC machine tools.

CAD was introduced in 1964, when IBM made it commercially available. The first complete (turnkey) system was made available in 1970 by Applicon Incorporated. Only recently, however, has the dramatic impact of this new technology been felt. By the mid-1980s, CAD systems had become quite commonplace in manufacturing.

THE CAD SYSTEM

Various combinations of equipment go to make up a CAD system. This holds true for small, medium-sized, and large system applications. The specific package selected depends largely on the needs of the user. Various types of drawings, referred to as **hard copy,** may be preferred by some, while others need no hard copy.

CAD systems are summarized as follows:

1. CAD systems are categorized as micro, mini, and mainframe.

2. Equipment for *every* CAD system will include processing, input, and output.

3. Processing includes programs magnetically stored on media and the means to drive them. The three types of media are cartridge, floppy disk, and hard disk.

4. Common input equipment includes keyboard, digitizer/puck or light pen, and mouse.

5. Output equipment is categorized as CRT, plotter, printer, and CAM.

The state of the art in CAD is still evolving. Future systems will enable the user to "talk" to the computer. The ultimate goal of CAD includes a full implementation with CAM and a totally automated factory.

HARDWARE

CAD systems require hardware (i.e., a computer and associated devices such as terminals and plotters) and software (i.e., program of instruction that makes the hardware operate in the desired manner). What follows is a brief discussion of some of the equipment that makes up a CAD system. It is not our intention to get into a long, drawn-out description of each piece of equipment but merely to familiarize you with some of the terms.

A **computer system** consists of three main areas of data handling—input, processing, and output—and is backed by a fourth, storage. These four components are shown in Figure 11-1.

Figure 11-1
Computer system components.

Input devices take data in machine-readable form and send it to the processing unit. The processor more formally known as the central processing unit (CPU) manipulates input data into the desired information by executing computer instructions. It has a memory that temporarily holds data and instructions (programs). Output devices make the processed information available for use. Storage devices are auxiliary units outside the CPU that store additional data and programs.

CAD systems are categorized by type of computer. A **microcomputer** is a small, low-cost computer built around a microprocessor. These computers have a single printed circuit board containing primary system chips. They are excellent for graphic applications where low price and small size are important. **Minicomputers** are smaller versions of large mainframe computers. Minicomputers have many of the capabilities of mainframe computers. The number of terminals that can be supported by a minicomputer varies from two to eight terminals. The large **mainframe** computer is distinguished by its capacity, function, and cost. Its main memory capacity is several orders of magnitude larger than the minicomputer and its computational speed is several times that of a mini- or microcomputer. In general, the larger the computer, the more sophisticated the CAD system and the higher the cost.

The **workstation** is the vehicle for communication with the CAD system. It represents a significant factor in determining how convenient and efficient it is for a designer to use the CAD system. The workstation must accomplish five functions:

1. Interface with the central processing unit

2. Generate a steady graphic image for the user

3. Provide digital descriptions of the graphic image

4. Translate computer commands into operating functions

5. Facilitate communication between the user and the system

The use of interactive graphics has been found to be the best approach to accomplish these functions. A typical interactive graphics workstation consists of a graphics terminal, operator input devices,

Figure 11-2
Workstation showing terminal,
keyboard, and mouse.

ᴬⁿᵈ sometimes output devices. A graphics design workstation showing two of these components is shown in Figure 11-2. A graphics terminal consists of a television screen [**cathode ray tube** (CRT)] and a keyboard. The CRT is the most crucial device in the system. By displaying letters and graphic representations of geometric shapes and design on the screen, the computer communicates with its human partner. In turn, he or she can "talk back" by using various input devices. This activity is possible because the CRT (also called the *monitor*) and input devices are connected directly to the computer.

The alphanumeric **keyboard** is one of the primary input devices. The keyboard is mainly used for text and to increase the precision of dimensioning. It is the most common means of communicating with the computer when operating outside the CAD mode. Figure 11-2 also shows a typical keyboard.

A **digitizer** is an electromechanical input device that resembles an electronic tablet and is shown in Figure 11-3. It allows direct input to a computer using data in the form of digital coordinates. The surface of the tablet corresponds to the face of the CRT. A puck or light pen is used to transmit information to the computer. The digitizer or tablet is

Figure 11-3
Digitizer/puck with menu.

suitable for interaction since it allows the designer to work naturally with the puck or light pen and simulate freehand sketching. Often the tablet is divided into areas with significant meaning to the drafting system. These areas are called a **menu.** In effect, the menu gives access to a specific instruction or small drawing routine in the computer system. The use of the menu eliminates the need for typing coded instruction and simplifies the CAD system.

A command menu instructs the computer what to do, but it does not always specify where on the display to execute the instruction. To specify this position, a **mouse** is used to move a cursor to any position on the face of the screen. Note that the puck, light pen, or keyboard can also be used to move the cursor. The mouse, shown in Figure 11-2, has a ball on its underside which rotates when the device is pushed along a surface. As the ball or wheels rotate, the cursor position on the graphics screen is determined. The mouse has buttons that provide for additional communication with the computer.

The **plotter** is the most common hard-copy output device and is the peripheral that produces the finished drawing. There are several types of plotters. They are drum, flatbed, vertical bed, photo plotter, electrostatic printer/plotter, and ink jet plotter. One of the most common is the drum plotter, and it is shown in Figure 11-4. Drum plotters have paper partially wrapped around a drum which rotates, giving longitudinal paper movement under a pen carriage. The pen carriage traverses parallel to the axis of the drum, producing lateral pen movement.

A flatbed plotter is shown in Figure 11-5. They are used when increased accuracy is desirable. As the name implies, flatbed plotters operate by plotting on paper that rests on a flatbed. The pen moves in the X and Y directions across the bed. Vertical bed plotters, or easels, are an alternative to the flatbed type. This type of plotter is like a flatbed rotated through 90° so that it assumes the position of a conventional drawing board. Flatbed plotters combine high speed with accuracy.

Another output device is the **printer,** and one is shown in Figure 11-6. Printers are much faster than plotters; however, they are not as accurate. Dot-matrix printers are relatively inexpensive, but line quality is not as good as a plotter. Laser printers have much better line quality; however, they are much more expensive.

**Figure 11-4
Drum plotter.**

Figure 11-5
Flatbed plotter.

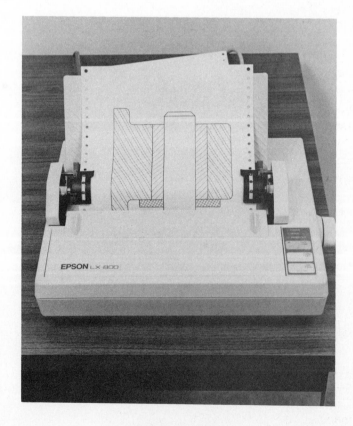

Figure 11-6
Dot-matrix printer.

SOFTWARE

CAD hardware, or for that matter, any computer hardware would not be of much use unless there was suitable software to support it. Without such software the computer cannot function, "like a stereo system without any tapes." The term "software" is used to refer to all programs that run on a computer system to perform various tasks. In general, CAD software can be divided into three main categories: operating system, graphics system, and application system.

The **operating software** is the major system in a CAD system. It provides for such tasks as memory allocation, the scheduling of the processing unit, the operation of the input/output devices, and it arranges the priorities of operations. Most CAD systems have some form of auxiliary storage device, usually a disk drive, and one section of the operating system has to oversee the disk storage system. This part of the operating system is called the *disk operating system* (DOS) and it has the task of organizing the transfer of files between disk and memory as well as keeping a check on the location of files. The operating system software also acts as an interface between the user and the computer. This interface enables the user to communicate with the computer via the operating system language.

The function of the **graphics software** is to provide graphics capabilities. To meet the requirements of designers and drafters, graphics software must have the capability of accepting, editing, revising, correcting, and storing data for primary graphic entities such as points, lines, planes, arcs, and circles as they combine to form multiview and pictorial representations.

Application software is software created for specific tasks to be performed in one of the technical fields, such as architectural, mechanical, structural, electronic, and highway design. It is the application software that turns the CAD hardware into a useful system. There are many different commercially available systems. CADKEY, AUTOCAD, and VERSACAD are a few of the more common ones.

COMPUTER-AIDED DRAWING

Today a drawing may be made by using one or both of two basic techniques: instrument drawing or CAD. We have covered the basic fundamentals of drawing, from the aspect of sketching. Points, lines, and surfaces have been used to describe objects by means of orthographic views, pictorials, sectional views, auxiliary views, tolerancing, and dimensioning. These fundamentals are essential for CAD.

It is at this point that one should be in a position to understand the unique coming together of a human being and a CAD system. The human being directs, provides the objectives, and makes the final evaluation. The CAD system performs the operations by supplying information and instructions as to how, when, and where operations are to be carried out.

In the modern CAD system the drawings are constructed interactively by the user supplying information to a computer via commands or with a menu-driven series of instructions. The drawings can easily be edited, manipulated, and stored for future retrieval. A plotter or printer is used to produce the final drawing to high standards of accuracy and line consistency.

INPUT METHODS

The keyboard is used primarily to input textual material, dimensions, and operating system commands. However, some of the applications software packages are designed to create the entire drawing with only keyboard entry. This method is not as efficient for geometric shapes as other input methods.

Some CAD systems use a large digitizer which resembles a traditional drawing board. Rough sketches are taped in place on the digitizer and traced electronically using a light pen or puck. Lines, arcs, circles, and other geometric shapes and forms are converted into electrical impulses and recorded as digital data in the computer's memory. The computer tidies up the drawing by straightening lines, smoothing out curves and arcs, and orienting lines at the proper angles. If desired, these data may be called up and displayed on the CRT screen, where any necessary manipulations can take place.

In one form the digitizing tablet has an electronic grid system. Generators acting within the tablet produce discrete signals in response to the puck or light pen. A cross-hair symbol, called a cursor, that corresponds to the light-pen position will appear on the CRT screen. It is by this means that x-y coordinates are created and stored in the computer's memory. Thus a drawing becomes a series of x-y coordinates.

Another input method is the mouse, which controls the cursor on the screen. A mouse has the same capability as a digitizer/puck or light pen. However, it operates on any surface, with the relative position of the mouse ball determining the position of the cursor.

All of the many commands needed to communicate with most CAD systems are stored in memory and activated by a menu. Menus are displayed either on the screen or on the digitizer, as shown in Figure 11-3. Selection of the desired command is made by a variety of methods, such as keyboard, digitizer/puck or light pen, or mouse. Menus are hierarchically structured, as shown in Figure 11-7. To create a circle in this particular CAD system, you would first activate the CREATE function from the main menu. Next, the CIRCLE command is activated from the CREATE menu. Finally, the method of constructing the circle is activated by one of the CIRCLE commands, such as CTR + DIA, which stands for "center and diameter." The center may be located by the cursor or keyed in with the x-y coordinates. Of course, the diameter must be keyed in. A similar process is followed to dimension the circle. Within a short period of time, the user becomes quite proficient in using menus. In addition, prompts are given to make most systems quite user-friendly.

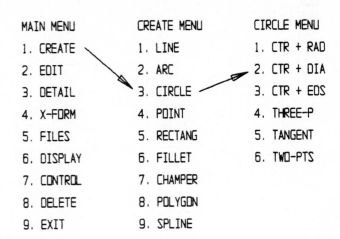

Figure 11-7
Hierarchically structured menu
for creating a circle.

CAD ENHANCEMENT

CAD can do much more than merely duplicate those activities associated with instrument drawings. The abilities of a CAD system include some or all of the following functions: zoom, mirror, rotate, move, copy, crosshatch, automatic dimension, and spatial coordinates.

Zoom is the capability of the system to either enlarge or reduce a portion of a drawing displayed on a CRT screen. Using the zoom function enables one to fill the screen with a specified portion of the drawing detail. This is a viewing facility that could be compared to having a powerful magnifying glass to examine fine details. The zoom activity does not modify the drawing in any way. Figure 11-8 shows a window/section of a drawing to reveal fine detail.

Figure 11-8
Zoom function.

(a) Original   (b) Mirror

**Figure 11-9
Mirror function.**

**Figure 11-10
Rotate function.**

(a) Original   (b) Copy

**Figure 11-11
Copy function.**

**Figure 11-12
Crosshatching.**

Objects that are symmetrical about an axis may be *mirrored;* that is, the operator may create one-half of a view of an object on the CRT and then automatically produce the second half. Figure 11-9 illustrates this feature.

The **rotate** function enables the user to turn a two- or three-dimensional part displayed on a CRT through a selected angle about an axis. Different views of the object as illustrated in Figure 11-10 can be automatically created on the screen. The designer or drafter can select the most appropriate one.

A CAD system also has the ability to **move** a portion of the display that has been positioned incorrectly on the screen. This function is accomplished by creating a window around the desired information and moving it to the new location by means of the cursor or by using coordinates.

The **copy** function saves the drafter considerable time when there are a number of objects or symbols that need to be repeated. Once the original has been created, the copies are positioned by the cursor or by coordinates. Figure 11-11 illustrates the copying of three sets of additional holes for the object. The copy function is similar to the move function.

It is often necessary to produce cross sections on an engineering drawing to reveal the interior details of a certain component. Cross sections are commonly indicated on a drawing by **crosshatching,** which if done manually is very time consuming and tedious. Most CAD systems can do this task automatically. The appearance of crosshatching varies depending on the hatch angle, the spacing between successive hatch lines, and the pattern. Figure 11-12 illustrates this variation in crosshatching.

Most CAD systems provide the operator with a complete range of automatic dimensioning conventions which cover the ANSI standards. The capability to create dimensions to a standard format is a prime requisite of a good CAD system. Dimensioning of a drawing on a CRT screen is done automatically or by the operator using a menu. In either case, extension lines, dimension lines, and arrowheads are supplied in accordance with good practice. Automatic dimensioning is possible because the object is defined in space by its XYZ coordinates.

ADVANTAGES

All of the enhancements given in the preceding section are also the advantages that CAD enjoys when compared to instrument drawings. In addition to those advantages, and to some extent because of them, CAD is faster, more accurate, more consistent, neater, easier to store, and easier to make corrections and revisions.

Once a CAD operator is trained, lettering and linework tasks are much faster than instrument drawings. CAD hardware is accurate to 0.001 inch, and this cannot be duplicated by the human drafter. Linework and lettering are the major strengths of CAD because the output is neat and consistent.

One of the major advantages of CAD is in better storage capability. Computer storage of drawings is much more space efficient than manual storage. Also, the deterioration problem associated with traditional drafting media is eliminated.

An additional advantage of CAD occurs when a drawing needs to be corrected or revised. All of the advantages present when a drawing is created are applicable to this activity.

A major advantage is the link with manufacturing. The information stored in the computer can be used to manufacture the part or assist in its manufacture.

CAD is one of the foundations of a paperless company. Access to any drawing is accomplished by direct linkage, by telephone lines, or by satellite to the database.

SUMMARY

In summary, one must know or do the following:

1. The term CAD/CAM stands for "computer-aided design and computer-aided manufacturing."

2. A computer system consists of three main areas of data handling—input, processing, and output—and is backed by a fourth, storage.

3. Computers are categorized as micro, mini, and mainframe.

4. A typical interactive graphics workstation consists of a terminal, input devices, and output device.

5. CAD hardware consists of CRT, keyboard, digitizer/puck or light pen, mouse, plotter, and printer.

6. The three main categories for CAD software are operating, applications, and graphics.

7. Input methods are digitizer/puck or light pen, mouse, and keyboard. These methods are used in conjunction with a menu.

8. CAD is enhanced by the following functions: zoom, mirror, rotate, move, copy, crosshatch, automatic dimension, and spatial coordinates.

9. In addition to the enhancement advantages of CAD, it is faster, more accurate, more consistent, neater, easier to store, and easier to correct or revise than instrument drawing. Also, the link with manufacturing is easily effected.

EXERCISES

Name: _____

Copy the two-view sketch. Copy the lamp at $1\frac{1}{2}$ times its given size.

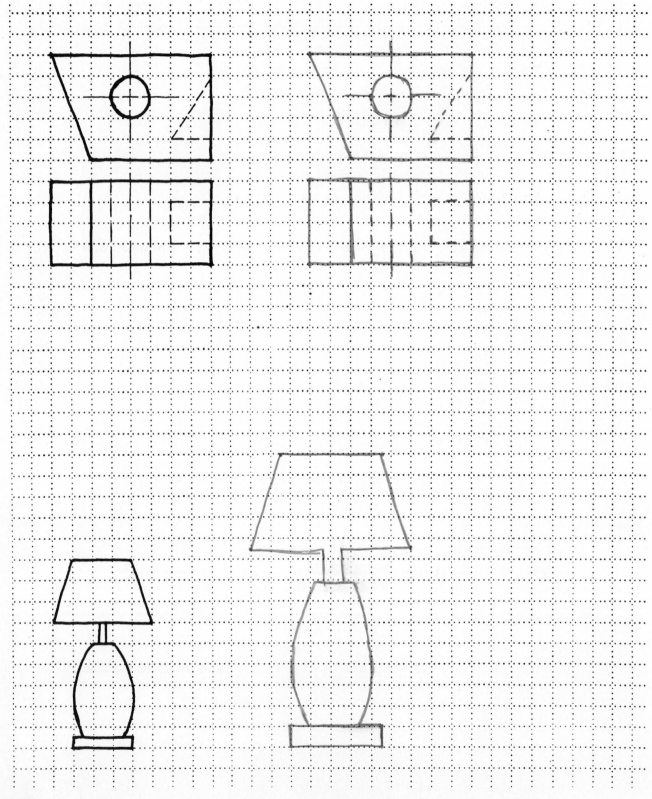

Circle the correct answer.

T (F)   **1.** It is easier to make the sketch if the paper is taped to the table.

(T) F   **2.** If an instrument drawing takes 8 hours to complete, a sketch of the same object might be completed in 1 hour.

(T) F   **3.** Sketches are used to create, explain, and record ideas.

T (F)   **4.** In sketching circles, the circle is sketched first and then the center lines.

T (F)   **5.** Sketches are drawn to scale.

T (F)   **6.** Center lines are the most prominent on the sketch.

(T) F   **7.** Cross-section paper is a decided help in sketching.

T (F)   **8.** The best type of sketching paper is plain white.

(T) F   **9.** Computer graphics will someday make the use of drawing instruments obsolete.

(T) F   **10.** With practice a sketched line can have almost the same appearance as a line drawn with instruments.

11. Which of the following lead grades is used for sketching?
   a. 3H.
   (**b.**) HB.
   c. 5B.
   d. FH.

12. The basic rule for sketching a straight line is to:
   a. use a ruler.
   b. push the pencil.
   c. press hard.
   (**d.**) pull the pencil.

13. The type of line that describes visible surfaces is called an/a:
   (**a.**) visible line.
   b. hidden line.
   c. centerline.
   d. phantom line.

14. The type of line that locates the center of a circle or arc is called an/a:
   a. visible line.
   b. hidden line.
   (**c.**) centerline.
   d. phantom line.

15. The type of line that alternates a short dash (about 3 mm) and a space (about 1 mm) is called an/a:
   a. visible line.   **c.** centerline.
   (**b.**) hidden line.   **d.** phantom line.

16. The best technique for sketching a line between two points is to:
    a. keep your eye on your thumb.
    b. keep your eye on the starting point.
    c. keep your eye on the pencil point.
    d. keep your eye on the end point.

17. On $8\frac{1}{2}$ by 11 inch paper, a tiny gear in a watch would most likely be sketched:
    a. half size.
    b. full size.
    c. double size.

18. On $8\frac{1}{2}$ by 11 inch paper, a 10-inch frying pan would most likely be sketched:
    a. half size.
    b. full size.
    c. double size.

19. On $8\frac{1}{2}$ by 11 inch paper, a typical door handle would most likely be sketched:
    a. half size.
    b. full size.
    c. double size.

Sketch a side view of your desk chair.

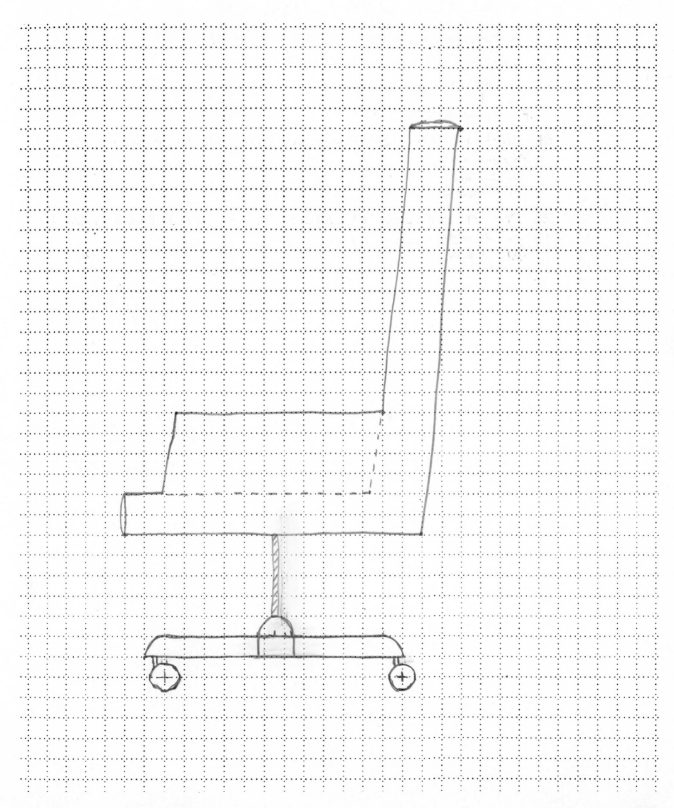

Circle the correct answer.

T F   **1.** The space between letters in a word should be the same for all letters.

T F   **2.** The vertical space between lines of lettering is the same as the height of the letters.

T F   **3.** The crossbar is below center in the letter **A.**

T **F**   **4.** The letter **M** would be a good example to use in illustrating stability of letters.

T **F**   **5.** Numerals are made twice as high as letters.

T **F**   **6.** Guidelines are made with an HB grade lead.

T F   **7.** Strokes are made from top to bottom and left to right.

T F   **8.** Stability is an important factor in the numerals 3 and 8.

T **F**   **9.** The usual height of letters is 10 mm.

T F   **10.** An HB-grade lead is used for lettering.

T **F**   **11.** Letters and numerals fall into two groups: straight line and inclined line.

T **F**   **12.** The widest letter of the alphabet is **X.**

T F   **13.** The space between words in a sentence is equal to the letter **O.**

Repeat each letter and numeral three times. Copy each sentence two times.

A A A A   I I I I I   N N N N   X X X X

E E E E   K K K K   T T T T T   Y Y Y Y

F F F F F   L L L L L   V V V V   Z Z Z Z Z

H H H H H   M M M M M   W W W W   B B B B

C C C C C   O O O O O   S S S S   3 3 3 3

D D D D D   P P P P P   U U U U U   4 4 4 4

G G G G G   Q Q Q Q Q   0 0 0 0   5 5 5 5

J J J J J   R R R R R   2 2 2 2   6 6 6 6

7 7 7 7   8 8 8 8 8   9 9 9 9

A PERSON CAN LEARN TO LETTER

A PERSON CAN LEARN TO LETTER

A PERSON CAN LEARN TO LETTER

TOP TO BOTTOM LEFT TO RIGHT

TOP TO BOTTOM LEFT TO RIGHT

TOP TO BOTTOM LEFT TO RIGHT

Copy each sentence two times.

THE METRIC SYSTEM IS USED

THE METRIC SYSTEM IS USED

THE METRIC SYSTEM IS USED

FOLLOW THE ORDER OF STROKES

FOLLOW THE ORDER OF STROKES

FOLLOW THE ORDER OF STROKES

PULL THE PENCIL, DON'T PUSH

PULL THE PENCIL, DON'T PUSH

PULL THE PENCIL, DON'T PUSH

STABILITY IS AN IMPORTANT FACTOR

STABILITY IS AN IMPORTANT FACTOR

STABILITY IS AN IMPORTANT FACTOR

THE GRID SERVES AS GUIDELINES

THE GRID SERVES AS GUIDELINES

THE GRID SERVES AS GUIDELINES

Sketch the three orthographic views of the objects. Label folding lines. Use grid to obtain proportion.

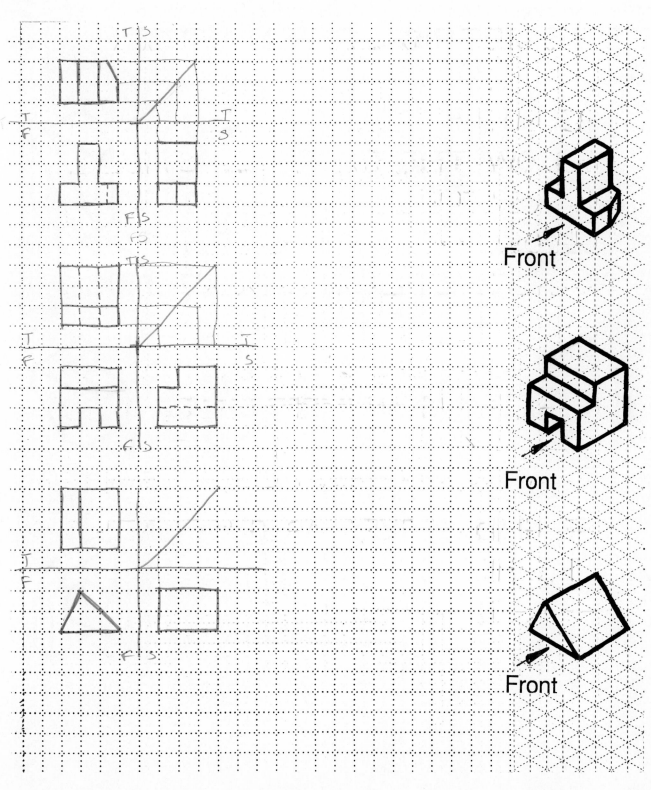

Front

Front

Front

Name: _____

Sketch the three orthographic views of the objects. Label folding lines. Use grid to obtain proportion.

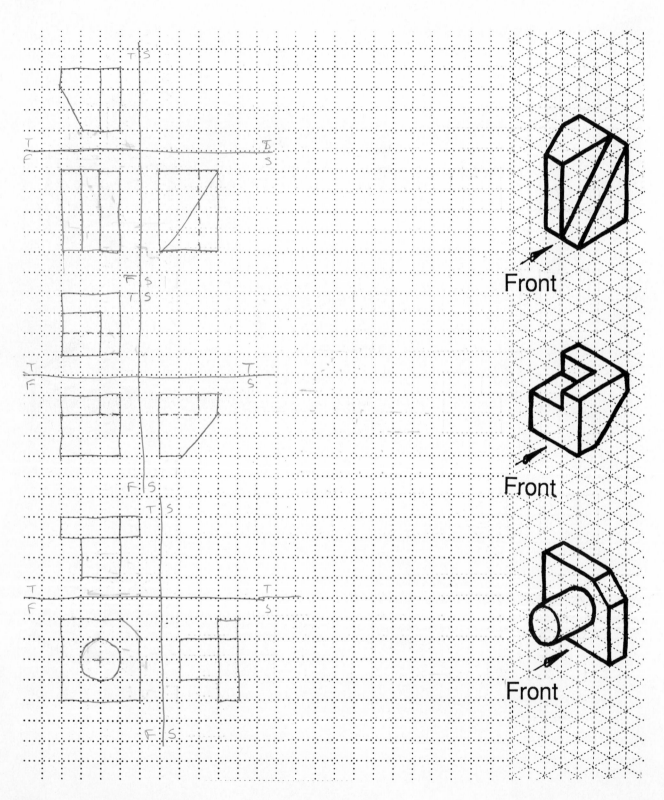

Front

Front

Front

EXERCISE 3-3

Name: _____

Sketch the three orthographic views of the objects. Label folding lines. Use grid to obtain proportion. The holes go through the objects.

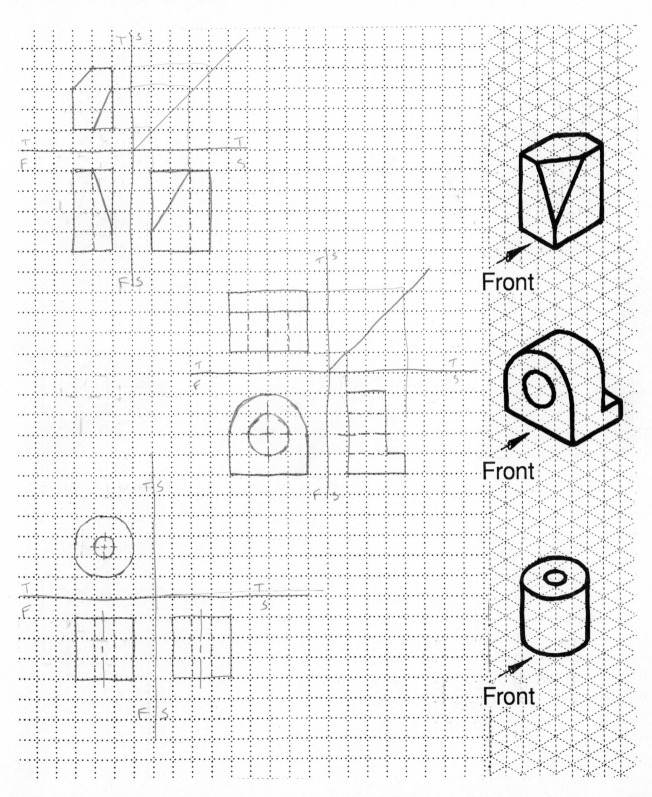

Front

Front

Front

EXERCISE 3-4

Name: _____

Sketch the three orthographic views of the objects. Label folding lines. Use grid to obtain proportion. The holes go halfway through the objects and are made by a drill.

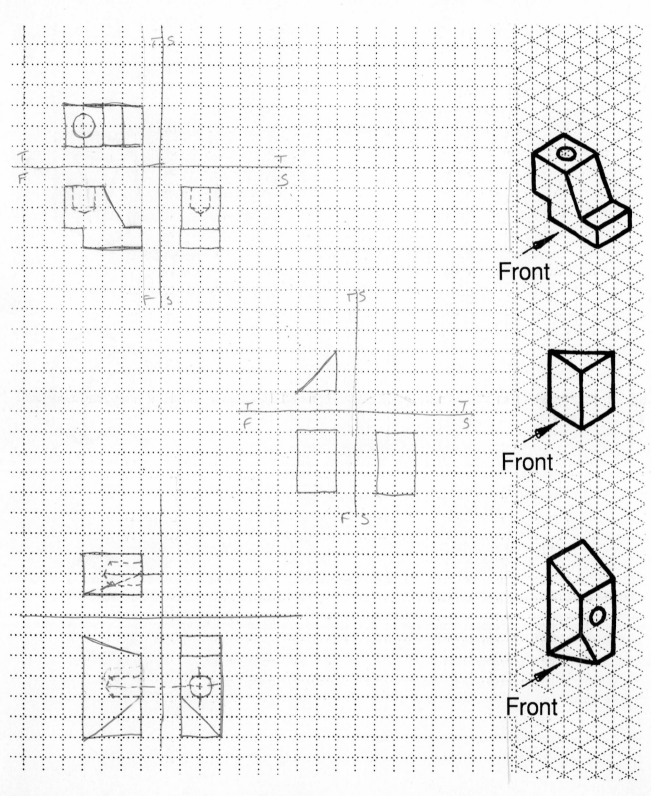

Front

Front

Front

Sketch the three-views of the normal lines of Figure 3-22a and b. The line is 25 mm long. Each square is 5 mm.

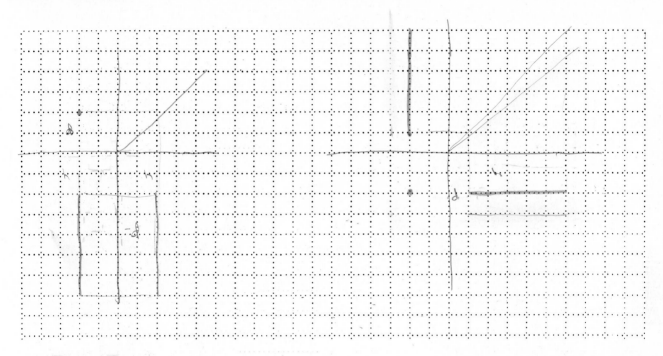

Sketch the three views of the inclined lines of Figure 3-25a and b. The line is 20 mm long and is at a 45° angle.

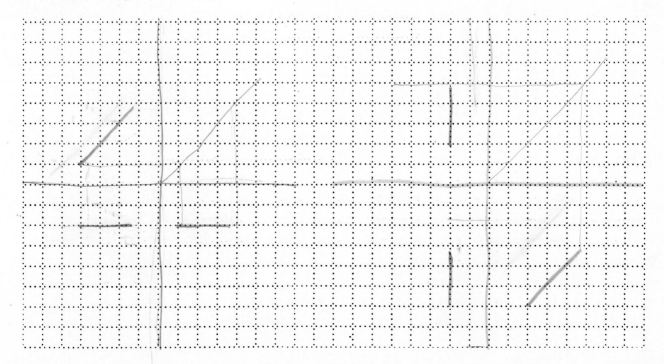

EXERCISE 3-6

Name: _____

Given two views of a normal surface, sketch the third. Use the 45° line where appropriate. Number all points. Label folding lines. The folding lines cross at the +.

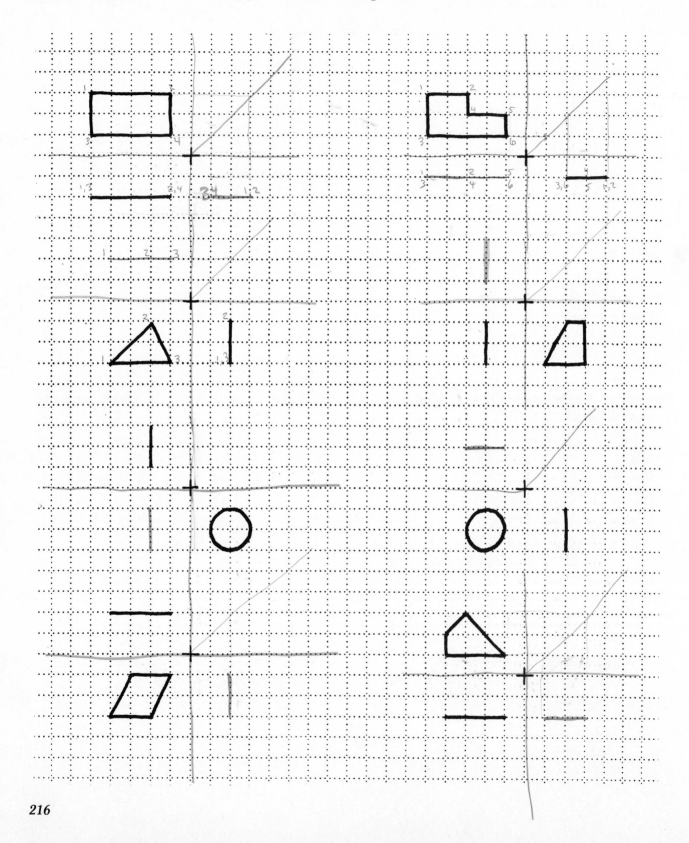

Given two views of an inclined or an oblique surface, sketch the third. Use the 45° line where appropriate. Number all points. Label folding lines. The folding lines cross at the +.

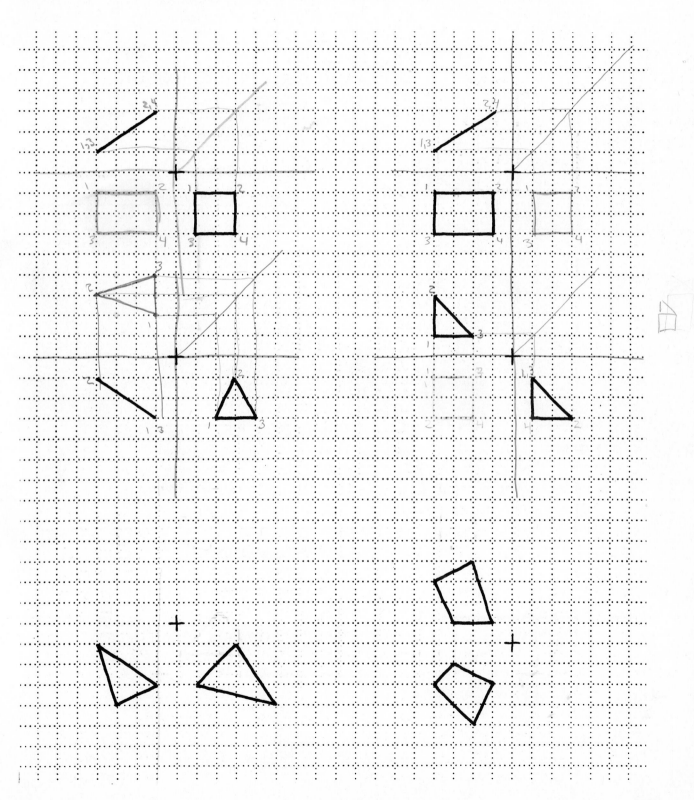

217

Name: _____

Identify the surfaces: normal, inclined, oblique, and cylindrical. Label the surface in the other
views as illustrated by the surface labeled "A."

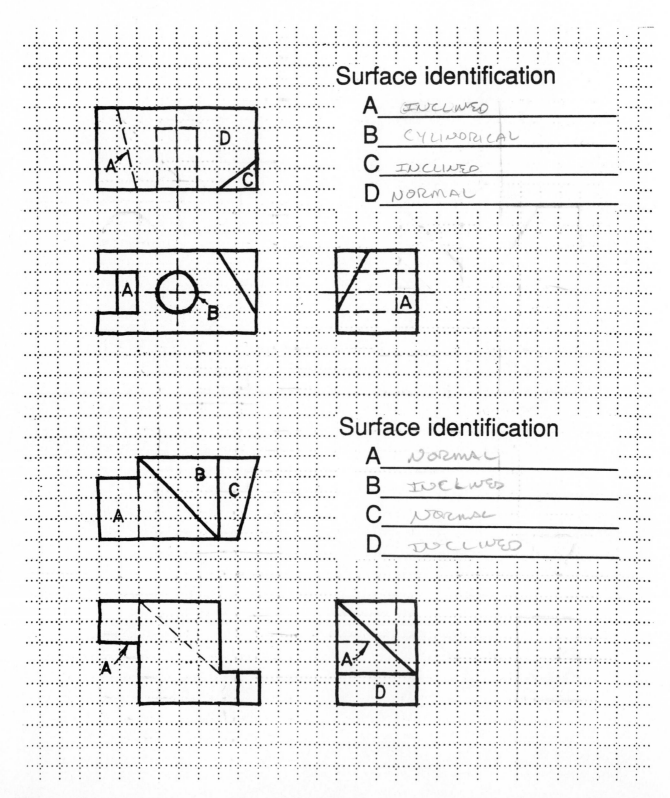

Surface identification

A   INCLINED

B   CYLINDRICAL

C   INCLINED

D   NORMAL

Surface identification

A   NORMAL

B   INCLINED

C   NORMAL

D   INCLINED

Given two views, sketch the third. Label folding lines. The folding lines cross at the +.

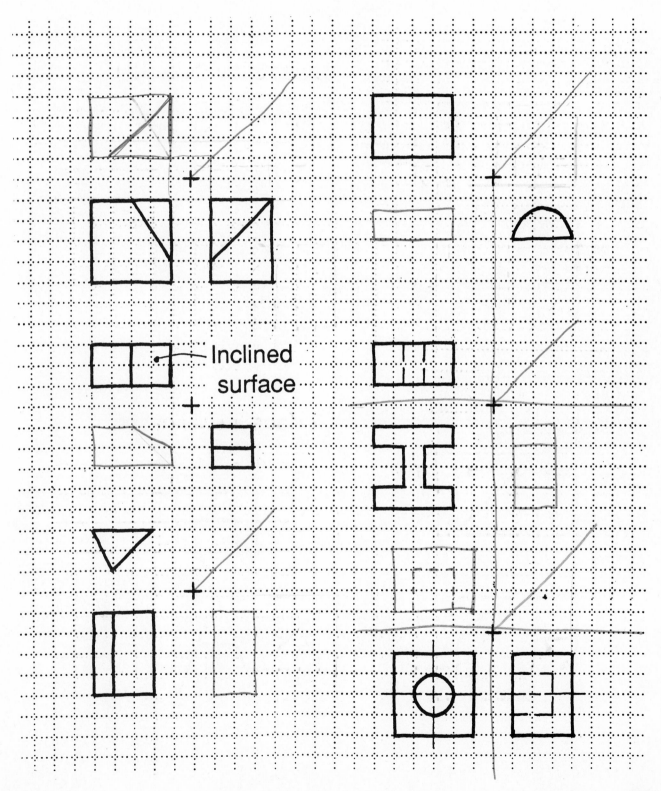

Inclined surface

Given two views, sketch the third. Label folding lines. The folding lines cross at the +.

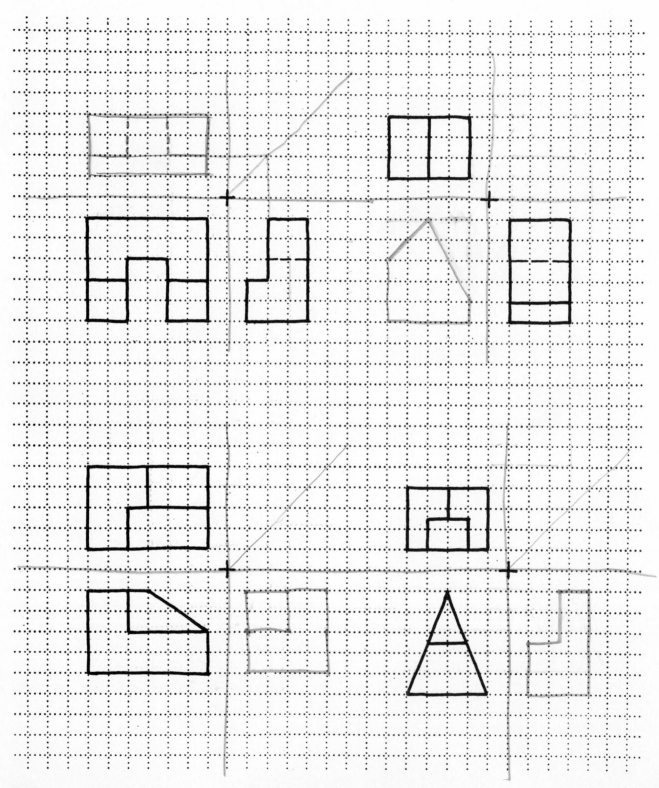

Given two views, sketch the third. Label folding lines. The folding lines cross at the +.

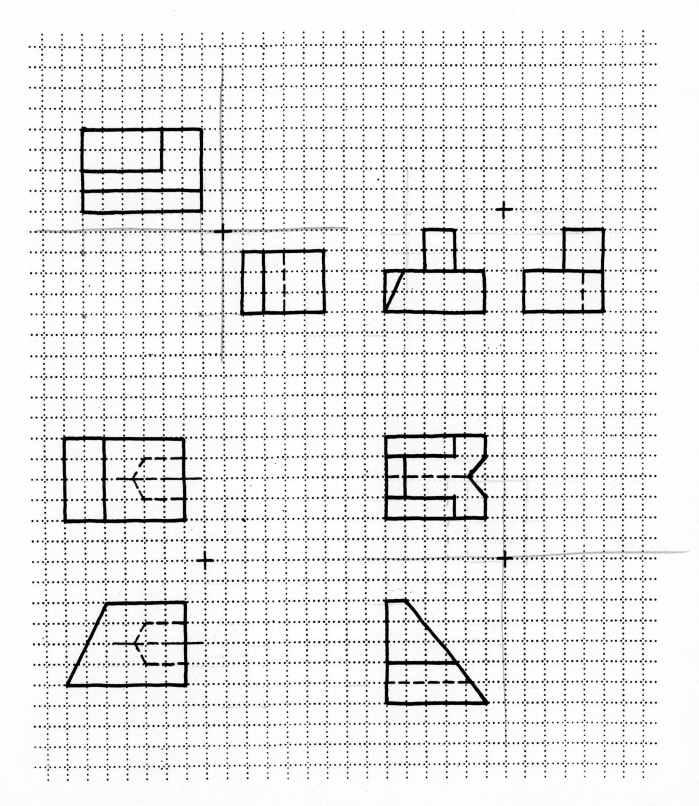

Given two views, sketch the third.

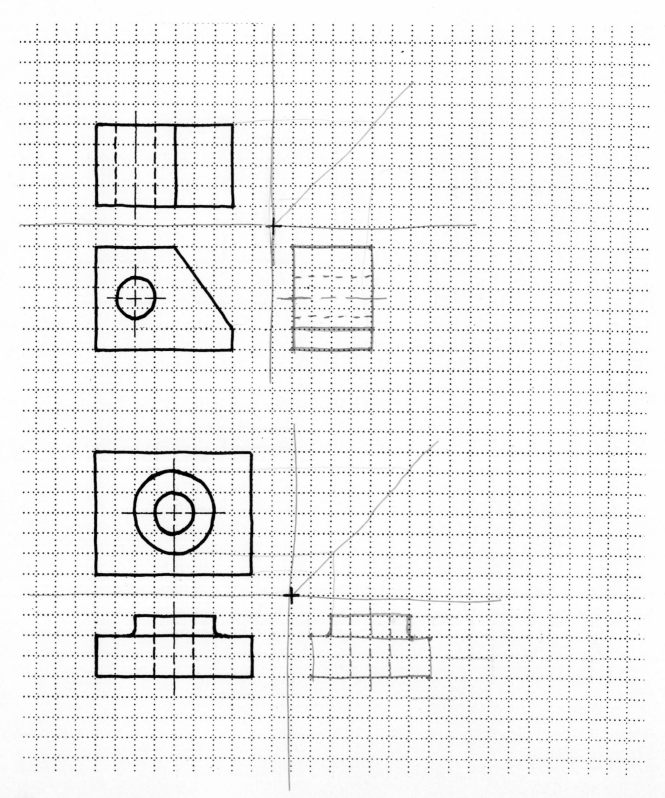

Name: _____

Given two views, sketch the third.

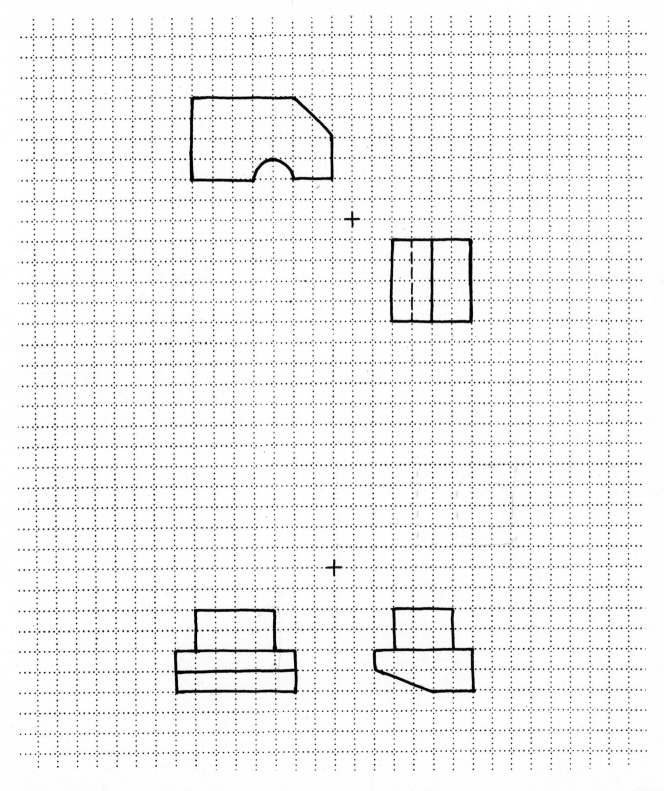

EXERCISE 3-14

Name: _____

Given two views, sketch the third.

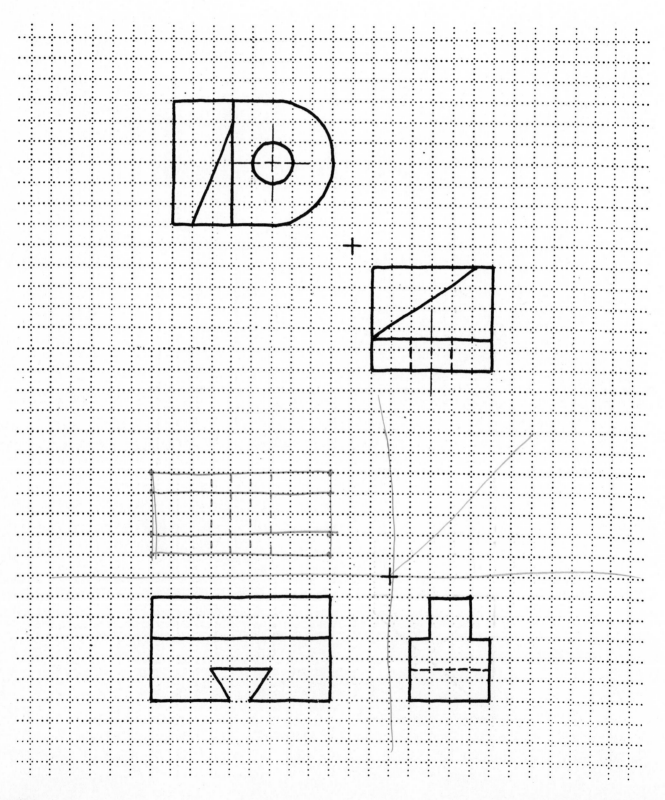

Given two views, sketch the third.

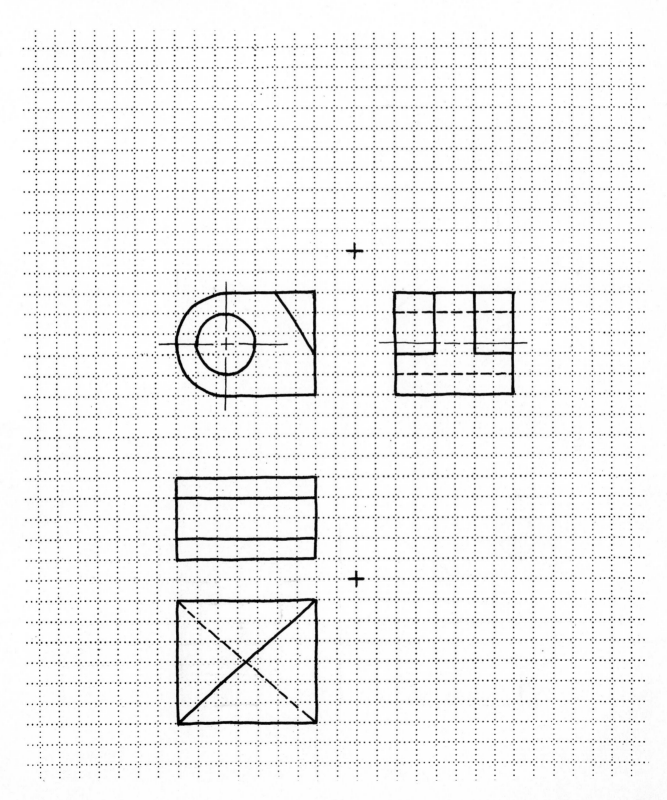

Name: _____

Given two views, sketch the third.

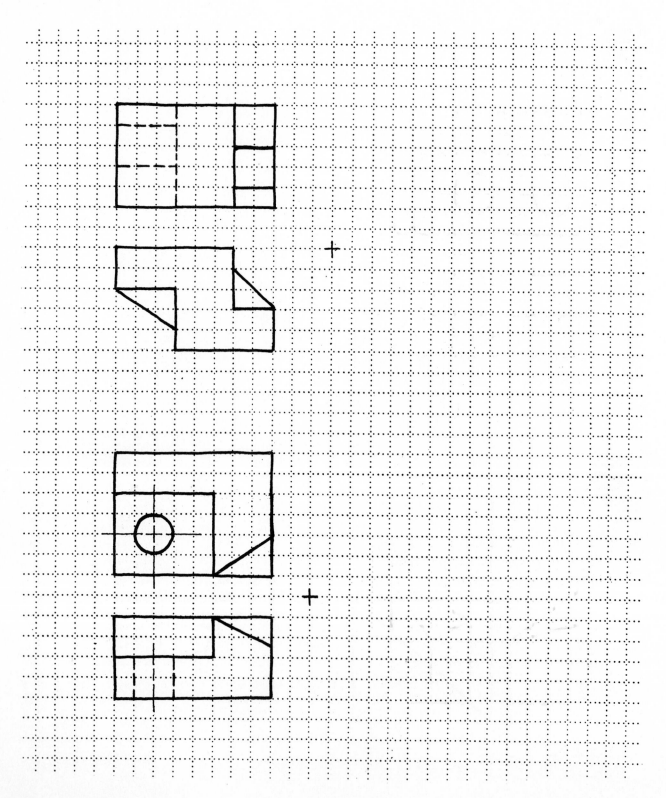

Name: _____

Given two views, sketch the third.

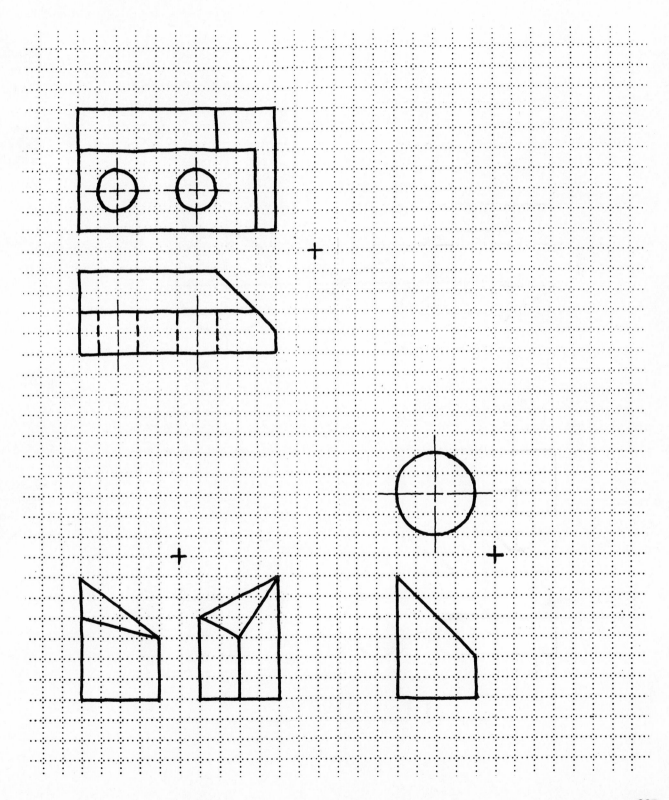

Given two views, sketch the third.

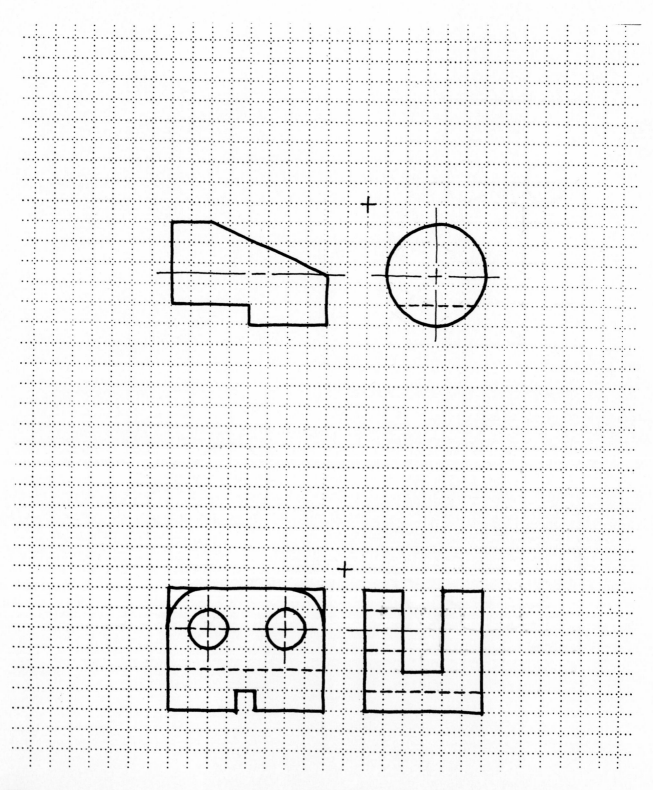

228

Given two views, sketch the third.

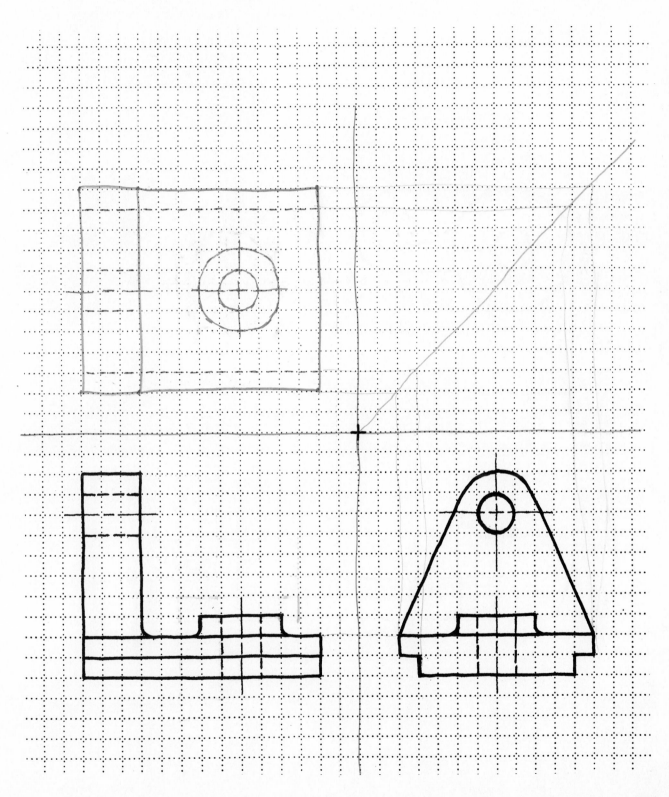

Given two views, sketch the third.

EXERCISE 3-21

Name: _____

Circle the correct answer.

T **(F)** 1. In orthographic projection, the height of an object is shown in the top view only.

(T) F 2. More than one view may be seen at one time when viewing an orthographic projection.

T **(F)** 3. The top and front views can describe any object.

(T) F 4. Three related views are usually necessary to describe an object.

T **(F)** 5. Hidden lines take precedence over visible lines.

(T) F 6. The top view shows only one dimension of the object.

(T) F 7. Orthographic projection is the basis of multiview sketches.

T **(F)** 8. In orthographic projection, the top and front views are generally placed in line horizontally.

(T) F 9. Looking perpendicularly at the front plane, the top and side projection plane appear as edges.

(T) F 10. The top view is a view looking down on the object.

T **(F)** 11. A projector is drawn with the same line weight as an object line.

(T) F 12. In a three-view sketch, three sets of projectors are used to produce the views.

T **(F)** 13. The top view is the principal view.

(T) F 14. In orthographic projection, the height of an object is shown in the front and side views.

T **(F)** 15. In orthographic projection, the width of an object is seen in the front and side views.

(T) F 16. Orthographic projection is right-angle projection.

(T) F 17. Projectors are lines of sight perpendicular to the projection plane.

T **(F)** 18. A point in space and the projection of a point in space are the same thing.

(T) F 19. Three views are required to describe most objects in orthographic projection.

(T) F 20. A hidden line takes precedence over a centerline.

(T) F 21. Hidden lines must be shown in all orthographic sketches.

T **(F)** 22. The angle, expressed in degrees, between adjacent projection planes is 120°.

T **(F)** 23. It is good practice to rotate the object for viewing when visualizing the different projections.

T **(F)** 24. A left-handed drafter should show a left-end view in preference to a right-end view.

(T) F 25. The view showing the most descriptive shape of the object should be selected as the front view.

(T) F 26. Two views will sometimes describe an object.

T F   27. Whether shown on the sketch or not, the projection planes are used.

T F   28. A numbering system can be helpful in projection points.

29. Orthographic projection requires that the projectors be:
    a. parallel to the plane of projection.
    b. perpendicular to the plane of projection.
    c. oblique to the plane of projection.
    d. None of the above.

30. Each view in orthographic projection shows only:
    a. one direction/dimension.
    b. two direction/dimension.
    c. three direction/dimension.
    d. four direction/dimension.

31. When looking at a three-view orthographic sketch, how many views can you actually see at one time?
    a. One.
    b. Two.
    c. Three.

32. The folding line between the top and front projection plane is labeled:
    a. T/F.
    b. F/S.
    c. T/S.

33. Which of the following lines is the most prominent on a sketch?
    a. Center.
    b. Right.
    c. Hidden.
    d. Visible.

34. The type of plane surface that is parallel to one of the projection planes is called:
    a. normal.
    b. oblique.
    c. parallel.
    d. inclined.

35. The type of plane surface that is perpendicular to one of the projection planes and inclined to the others is called:
    a. normal.
    b. oblique.
    c. parallel.
    d. inclined.

36. The type of plane surface that is neither parallel nor perpendicular to a projection plane is called:
    a. normal.
    b. oblique.
    c. parallel.
    d. inclined.

37. When viewing an object from the front, the side and top projection planes appear as a/an:
    a. perpendicular.
    b. cylinder.
    c. edge.
    d. surface.

38. A line that is parallel to two projection planes is called:
    a. normal.
    b. oblique.
    c. parallel.
    d. inclined.

39. A line that is parallel to one of the projection planes and inclined to the others is called:
    a. normal.
    b. oblique.
    c. parallel.
    d. inclined.

40. A line that is neither parallel nor perpendicular to any of the projection planes is called:
    a. normal.
    b. oblique.
    c. parallel.
    d. inclined.

Given the orthographic views of the two objects, sketch the isometric.

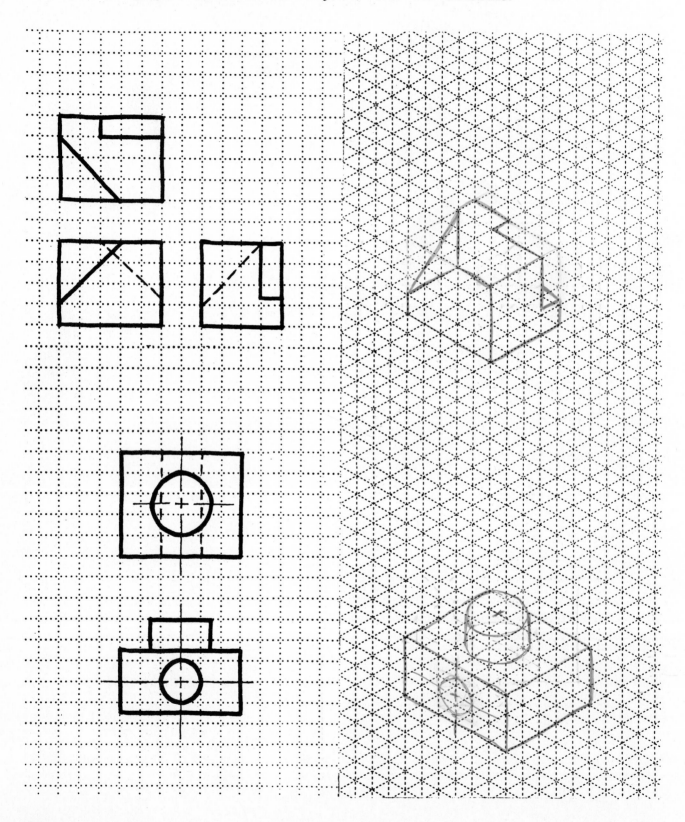

Given the orthographic views of the two objects, sketch the isometric.

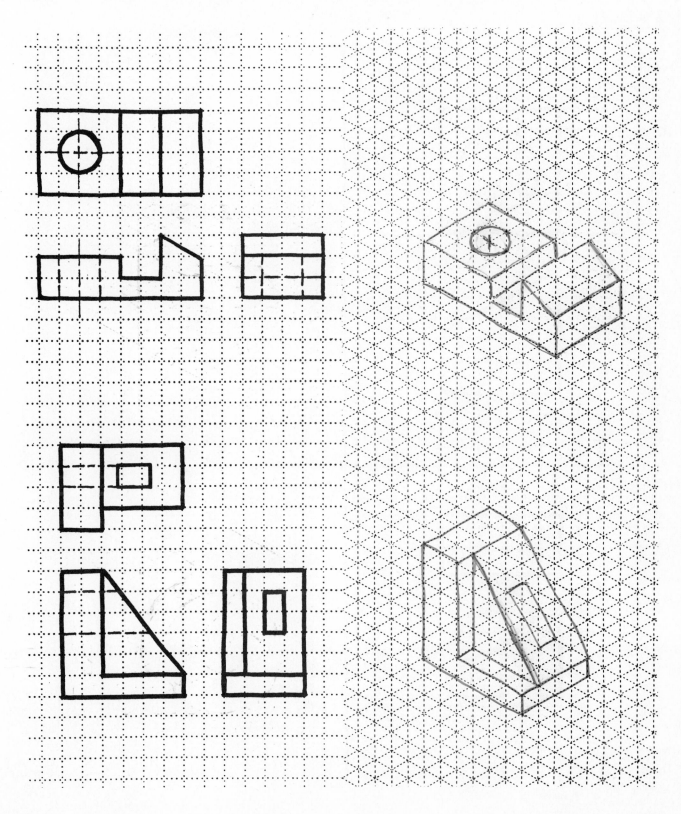

Given the orthographic views of the two objects, sketch the isometric.

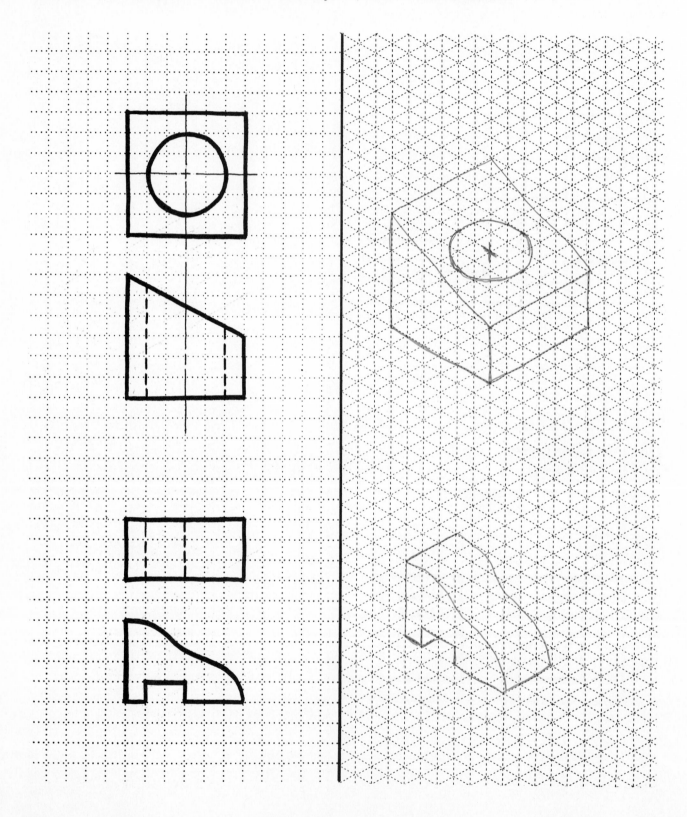

Construct an isometric sketch of a common object that has at least one inclined and one circular surface.

Given the orthographic views of the two objects, sketch the cabinet oblique.

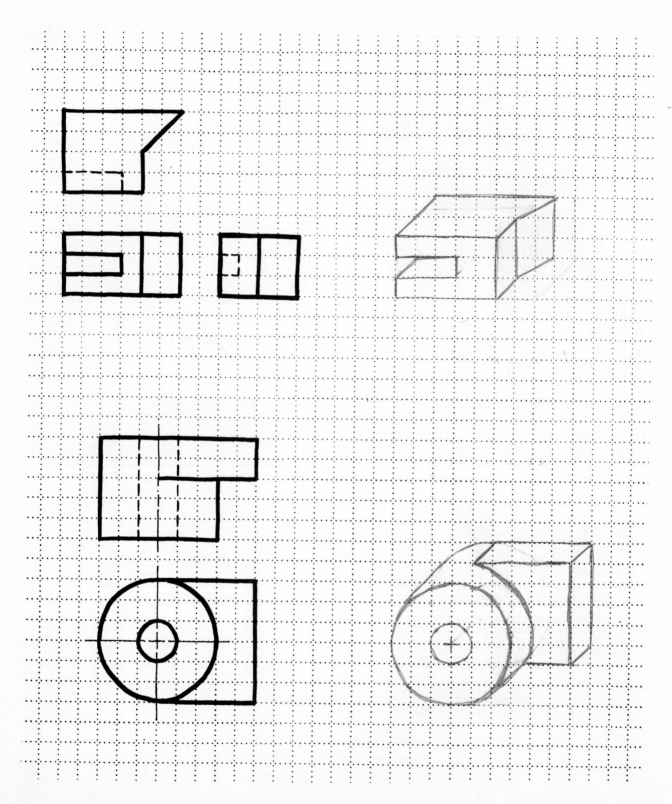

Given the orthographic views of the two objects, sketch the cabinet oblique.

Given the orthographic views of the object, sketch the cabinet oblique.

Construct a cabinet oblique sketch of a common object that has an irregular or circular surface.

EXERCISE 4-9   Name: _____

Circle the correct answer.

T **(F)**   **1.** In axonometric projection the object is in a stable position.

T **(F)**   **2.** Oblique sketching is a form of orthographic.

T **(F)**   **3.** Isometric projection uses three projection planes.

T **(F)**   **4.** The observer in orthographic and axonometric projection is at a finite distance.

(T) F   **5.** To construct circles in the isometric, a step-box method is recommended.

(T) F   **6.** In cabinet oblique sketches, the length along the 45° receding axis is reduced one-half.

(T) F   **7.** Circles will appear as true circles on receding surfaces of an oblique sketch.

(T) F   **8.** In isometric, irregular shapes require the plotting of a series of points with each point determined by three measurements.

T **(F)**   **9.** A true isometric projection is foreshortened by 75°.

(T) F   **10.** In axonometric projection, the parallel lines of sight are perpendicular to the projection plane.

T **(F)**   **11.** The most important sketching technique for an isometric is the square grid.

T **(F)**   **12.** Oblique sketching has the advantage that the front surface and all parallel surfaces are sketched the same as the front orthographic view.

T **(F)**   **13.** The position of the object in isometric should be accomplished to maximize the number of hidden lines.

T **(F)**   **14.** An orthographic square has two angles at 120° and two at 60°.

(T) F   **15.** In pictorial sketches, hidden lines are usually omitted.

T **(F)**   **16.** Orthographic projection uses only one projection plane.

(T) F   **17.** Circles will appear as ellipses on surfaces in isometric.

T **(F)**   **18.** Cavalier oblique sketches are recommended.

(T) F   **19.** Circular or irregular surfaces of the object should be selected for the front face of an oblique sketch.

(T) F   **20.** For isometric projection the object is rotated and tilted.

(T) F   **21.** Angles may not be measured by their degrees in isometric sketches.

(T) F   **22.** A reversed isometric axis occurs when we are looking up at an object.

(T) F   **23.** In an oblique sketch two axes are at 90° to each other.

T **(F)**   **24.** Trimetric projection is the most widely used because the axes angles and lengths are equal.

T **(F)**   **25.** Isometric is a form of oblique.

T **(F)**   **26.** In oblique projection the lines of sight are perpendicular to the projection plane.

T **(F)**   **27.** Oblique projection is a true projection.

28. In a cabinet oblique sketch:
    a. two axes are placed at 90° to each other, and the receding axis at 45° to one of the 90° axes.
    b. all axes are equally spaced at 120°.
    c. two axes are placed at 90° to each other, and the receding axis at 135° to one of the 90° axes.
    d. two axes are placed at 90° to each other, and the receding axis at 135° to one of the 90° axes and 45° to the other 90° axis.

29. Which of the following graphic presentations *cannot* be understood by the layman?
    a. Orthographic.
    b. Axonometric.
    c. Oblique.
    d. Perspective.

30. Vertical measurements in isometric can only be made on or parallel to the:
    a. vertical axis.
    b. inclined axis.
    c. horizontal axis.
    d. oblique axis.

31. In a cavalier oblique sketch:
    a. the receding axis is drawn at 45° and the length reduced one-half.
    b. the receding axis is drawn at 0° to 90° and the length on the receding axis reduced one-half.
    c. the receding axis is drawn at 45° and the length is full size.
    d. the receding axis is drawn at 0° to 90° and the length on the receding axis reduced one-half.

32. The relationship of the axes to each other in an isometric sketch is:
    a. 30°.
    b. 60°.
    c. 90°.
    d. 120°.

33. In a general oblique sketch, the receding axis is:
    a. perpendicular to one of the principal axes.
    b. parallel to one of the principal axes.
    c. at an angle of 0° to 90° with one of the principal axes.
    d. None of the above.

34. The type of axonometric projection that has two axes angles equal is called:
    a. dimetric.
    b. trimetric.
    c. isometric.
    d. cabinet.

35. One of the axes in an isometric sketch is *usually*:
    a. horizontal.
    b. parallel.
    c. vertical.
    d. None of the above.

Name: _____

Sketch full-sectional views of the objects. Label the cutting plane and sectional view.

SEC A-A

SEC B-B

Name: _____

Sketch full-sectional views of the objects. Label the cutting plane and sectional view.

SEC C-C

SEC D-D

Name: _____

Sketch full-sectional views of the objects. Label the cutting plane and sectional view.

Sketch half-sectional views of the objects. Label the cutting plane and sectional view.

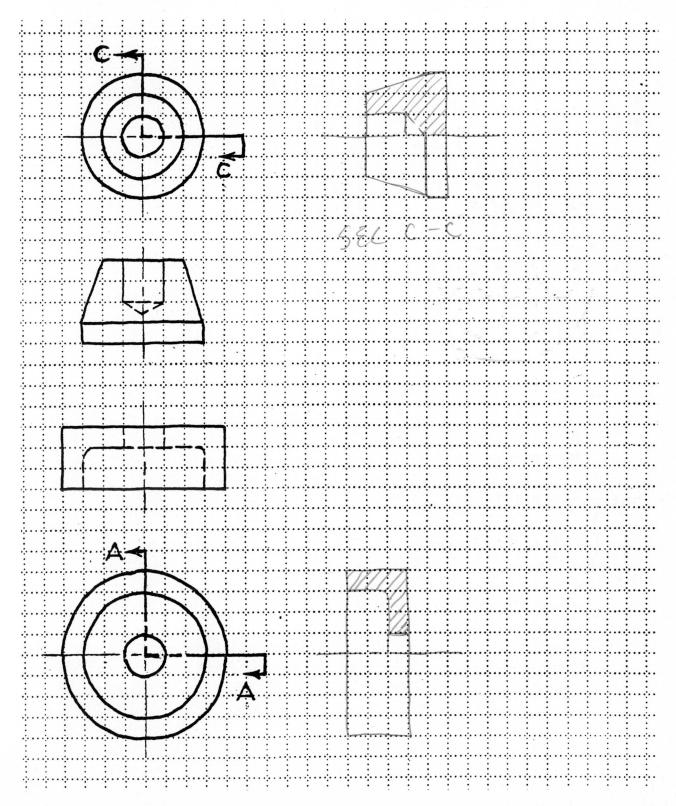

SEL C–C

Sketch half-sectional views of the objects. Label the cutting plane and sectional view.

Identify the different types of sectional views.

A OFFSET
B BROKEN / PARTIAL
C ASSEMBLY
D REVOLVED
E REMOVED
F FULL
   SECTIONAL

SEC B-B

SEC F-F

SEC A-A

Circle the correct answer.

T **(F)**   **1.** It is considered good practice to show hidden lines in the sectional view.

T **(F)**   **2.** The cutting plane for a half-section removes one-half of the object.

(T) F   **3.** Detail or removed sections may be drawn to an enlarged scale in order to emphasize detail.

(T) F   **4.** The angle most commonly used for drawing crosshatched lines is 45°.

(T) F   **5.** The crosshatched lines for the same piece in the same view on the sketch should slope the same angle and the same direction.

(T) F   **6.** "Conventional practice" is not a true projection. It is used because it makes the sketch easier to understand.

(T) F   **7.** When the cutting plane passes through and parallel to the axis of a bolt or shaft, it should not be crosshatched.

T **(F)**   **8.** In half-section views, hidden lines are generally included on unsectioned parts of the sectional view.

(T) F   **9.** A full section is used as part of an exterior view to show some detail without sketching a complete section.

(T) F   **10.** A cutting plane is an imaginary plane used to show the path of cutting an object to make a sectional view.

(T) F   **11.** A cutting plane line is made the same weight as a visible line.

(T) F   **12.** A cutting plane is passed through an object so as to reveal the shape or interior construction of the object.

(T) F   **13.** In a sectional view, only the material in contact with the cutting plane is crosshatched.

T **(F)**   **14.** A bolt or shaft is not crosshatched when the cutting plane passes perpendicular to its axis.

(T) **(F)**   **15.** A cutting plane line is shown in the sectional view.

(T) F   **16.** The first sectional view taken is labeled A-A.

(T) F   **17.** The arrows on a cutting plane point in the direction of sight.

T **(F)**   **18.** A section view is an exterior view with crosshatch lines added.

(T) F   **19.** A cutting plane indicates where the section has been taken.

T **(F)**   **20.** The letters on the cutting plane line indicate the direction of sight.

(T) F   **21.** It is poor practice to show a nonsymmetrical object in half section.

(T) F   **22.** A full-sectional view is an orthographic view.

(T) F   **23.** For a full section, the cutting plane is an imaginary plane passed entirely through the object.

T F   **24.** Crosshatched lines should not be drawn parallel to the contour lines of the object in the section view.

T F   **25.** All visible lines behind the cutting plane are shown in the sectional view.

T **F**   **26.** An individual part, when shown in a sectional view, would be classed as an assembly sketch.

T F   **27.** A cutting plane line takes precedence over a center line.

T **F**   **28.** A removed section is made directly on an exterior view.

T **F**   **29.** The cross section of a bar, spoke, rib, or other shape may be shown directly on the orthographic view by means of a detailed section.

30. Identify the following materials from their symbol.

a. _____TITANIUM_____   b. _____FABRIC, CORK_____

c. _____GENERAL_____   d. _____RUBBER_____

e. _____STEEL_____   f. _____BRONZE, COPPER_____

g. _____ELEC. WINDING_____   h. _____MAGNESIUM / ALUMINUM_____

i. _____ZINC_____

EXERCISE 6-1   Name: _____

Find the true length of the line. Label all folding lines, points, and true length lines.

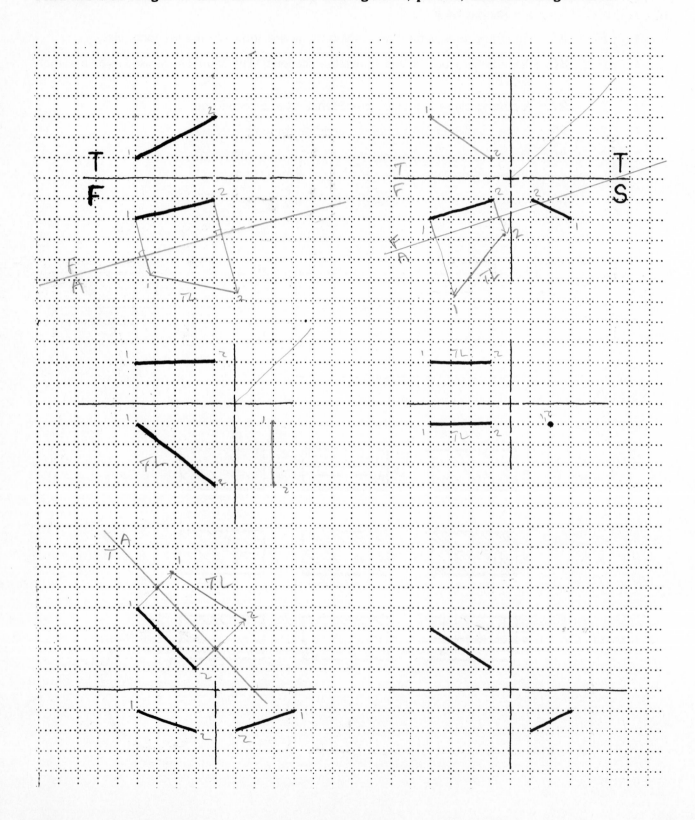

Find the true size of the inclined surface. Label the folding lines and true size surface. For the cylinder, use the axis as the folding line.

Name: _____

Construct partial auxiliary views for the inclined surfaces. The hole size is 10 mm.

Find the true size of the oblique surface. Take the first auxiliary off the top view in the upper figure and off the side view in the lower one. Use the squares to assist with the projection angle and the distance.

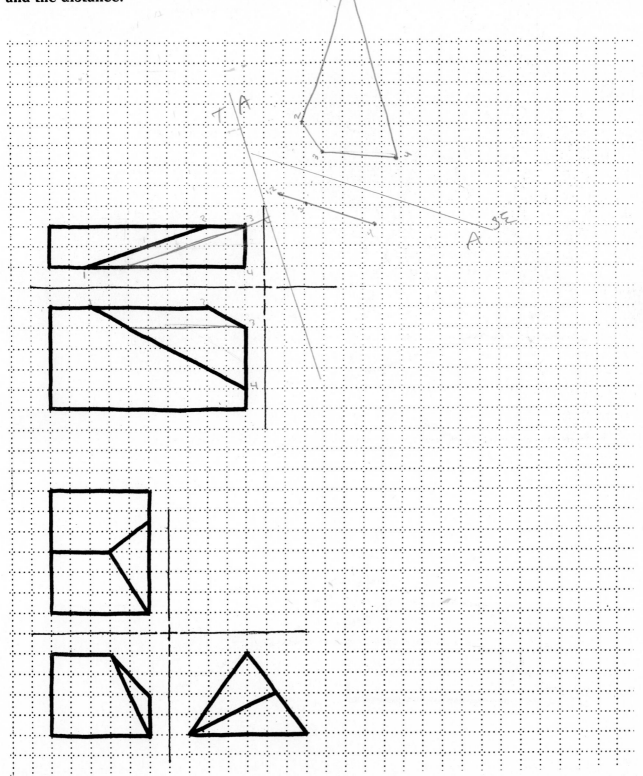

Name: _____

Find the true size of the oblique surface. Sketch a 10 mm square on the true-size surface and project the square back to the top and front view.

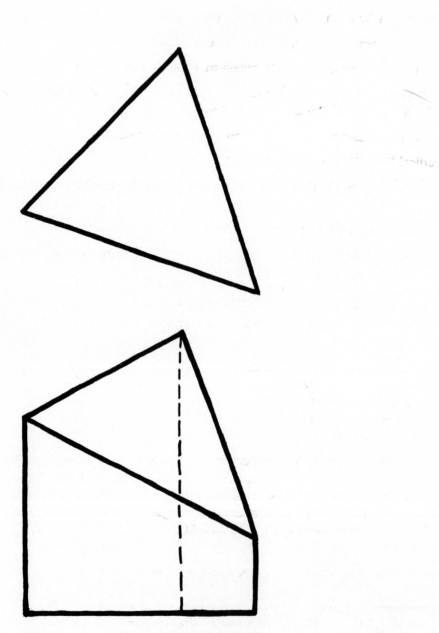

EXERCISE 6-6   Name: _____

Circle the correct answer.

(T) F   **1.** The three principal orthographic views are sometimes inadequate to fully describe an object.

T (F)   **2.** The auxiliary plane must in every case be placed parallel to the projection of the line or surface.

(T) F   **3.** An infinite number of primary auxiliary views taken off the top view are possible.

T (F)   **4.** Any line parallel to the front plane will show true length in the side view.

(T) (F)   **5.** An auxiliary view may be used to replace a primary view.

(T) (F)   **6.** Auxiliary views are used only to obtain true length and true size.

(T) F   **7.** The projection of an oblique line to an auxiliary view will show the line in true length provided that the folding line is parallel to the oblique line.

T (F)   **8.** It is always necessary to show the complete auxiliary view.

(T) F   **9.** A vertical line will appear as a point in the top view.

T (F)   **10.** A secondary auxiliary view is not necessary to obtain the true size of an oblique surface.

(T) F   **11.** Any level (horizontal) line will show true length in the top view.

12. To obtain the true size of an oblique surface, the secondary auxiliary folding line is placed:
  a. perpendicular to the edge view of the surface in the auxiliary view.
  (b.) parallel to the edge view of the surface in the auxiliary view.
  c. oblique to the edge view of the surface in the auxiliary view.
  d. None of the above.

13. Auxiliary views are used to find the:
  a. true length of an oblique line.
  b. true size of an inclined surface.
  c. true size of an oblique surface.
  (d.) All of the above.

14. When looking at the auxiliary projection plane taken off the front projection plane, the:
  a. projection plane appears as an edge.
  (b.) auxiliary projection plane appears as an edge.
  (c.) front projection plane appears as an edge.
  d. None of the above.

15. To obtain the true size of an oblique surface, the primary auxiliary folding line is placed:
  a. perpendicular to the top view.
  b. oblique to the true length of a line on that surface.
  (c.) perpendicular to the true length of a line on that surface.
  d. parallel to the true length of a line on that surface.

16. When a primary auxiliary plane is taken off the front projection plane, it is:
    a. perpendicular to the top projection plane.
    b. parallel to the front projection plane.
    c. perpendicular to the front projection plane.
    d. parallel to the top projection plane.

17. To obtain the true size of an inclined surface, the primary auxiliary projection plane is placed:
    a. parallel to the edge view of that surface.
    b. perpendicular to the edge view of that surface.
    c. inclined to the edge view of that surface.
    d. None of the above.

18. When a primary auxiliary view of an object is taken off the top view, the distance from a point l to folding line AT:
    a. is equal to the distance from point l in the front view to folding line T/F.
    b. is unequal to the distance from point l in the front view to folding line T/F.
    c. is equal to the distance from point l in the top view to folding line T/F.
    d. is unequal to the distance from point l in the top view to folding line T/F.

19. When looking at the auxiliary projection plane taken off the top projection plane, the:
    a. projection plane appears as an edge.
    b. auxiliary projection plane appears as an edge.
    c. top projection plane appears as an edge.
    d. None of the above.

20. What two types of auxiliary views are normally used?
    a. Principal and secondary.
    b. Alternate and primary.
    c. Principal and primary.
    d. Primary and secondary.

Dimension the objects. Each square is 5 mm. Indicate assumptions you make. All surfaces are finished.

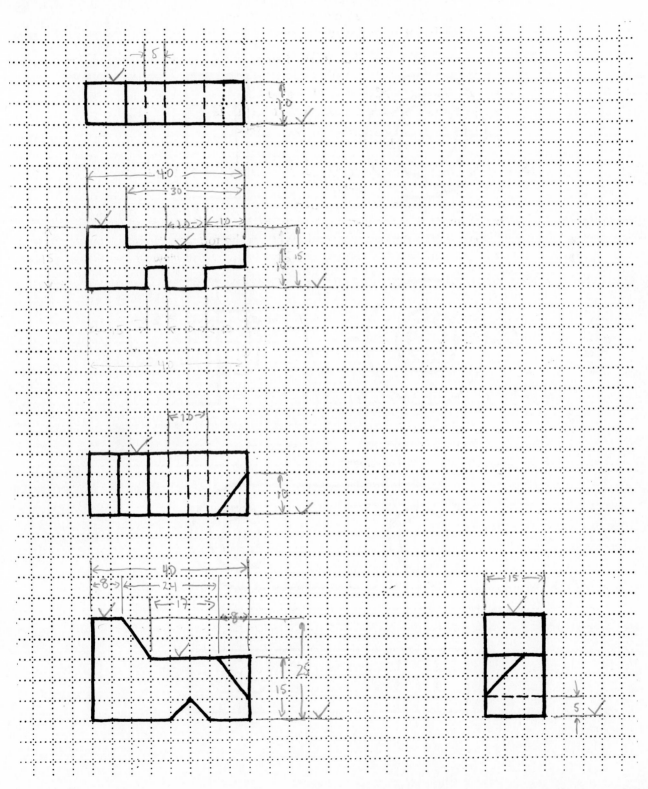

Dimension the objects. Each square is 5 mm. Indicate assumptions you make. All surfaces are finished.

Dimension the object. Each square is 5 mm. Indicate assumptions you make.

Name: _____

Dimension the object. Each square is 5 mm. Indicate assumptions you make. The object is a casting.

Dimension the objects. Each square is 5 mm. Indicate assumptions you make. All surfaces are finished.

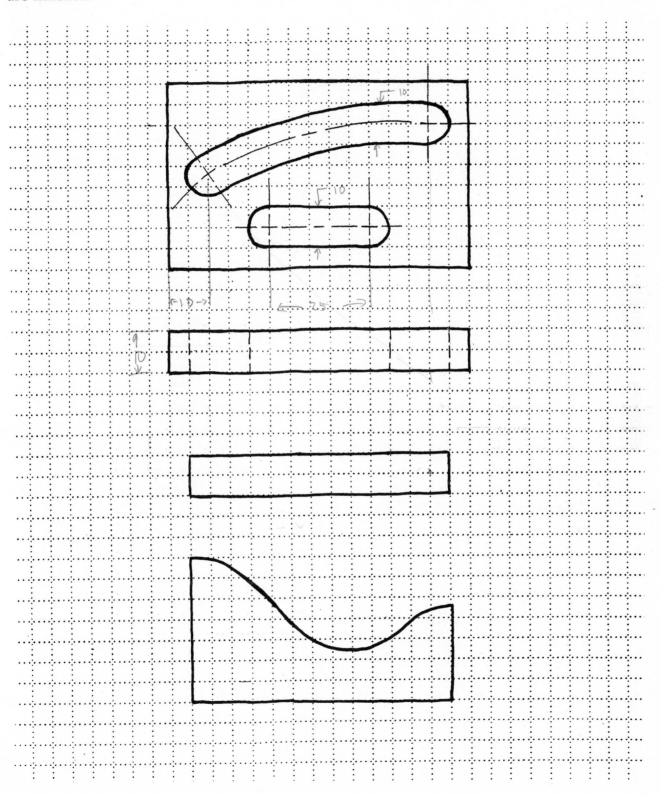

Dimension the objects. Each square is 5 mm. Indicate assumptions you make. The upper figure is a casting and the lower figure is finished all over (FAO).

Dimension the objects. Each square is 5 mm. Indicate assumptions you make. Use rectangular coordinates for the upper figure and polar coordinates for the lower. The holes are equally spaced in the lower figure and there is no counterbore.

Name: _____

Dimension the objects. Each square is 5 mm. Indicate assumptions you make. For the upper figure, use dimension lines. For the middle figure, omit dimension lines. For the lower figure, use the tabular form.

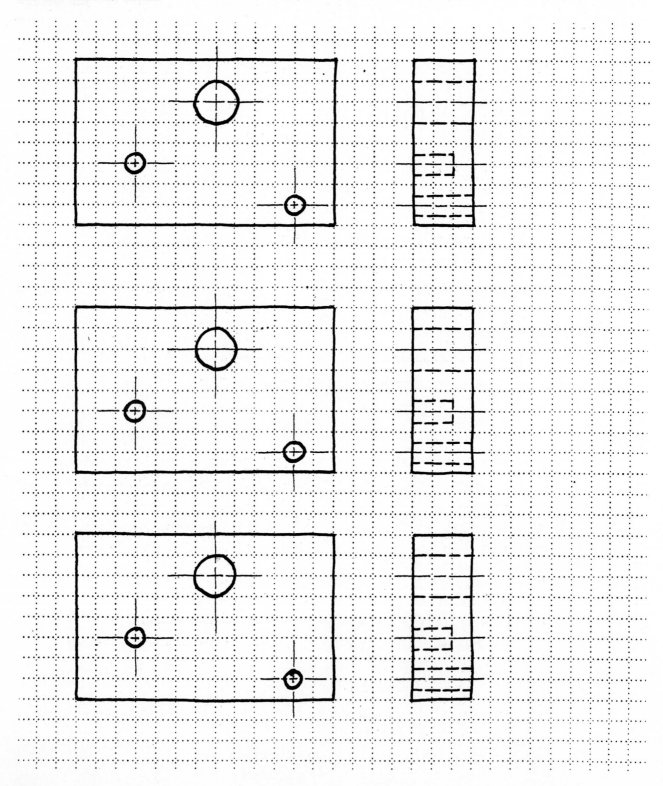

EXERCISE 7-9

Name: _____

Circle the correct answer.

T **(F)**   **1.** An extension line is a thin line with a break near its center for the dimension value.

T **(F)**   **2.** Extension lines should start 0.5 mm from the view and extend 3 mm beyond the last dimension line.

T **(F)**   **3.** As a general rule, when spacing dimensions from an object, the first dimension line should be 12 mm from the view, and all succeeding dimensions spaced 6 mm apart.

T **(F)**   **4.** Special dimensioning techniques are not required for small spaces.

(T) F   **5.** A center line may be extended and used as an extension line.

T **(F)**   **6.** Visible lines can always be used as extension lines.

(T) F   **7.** Dimension lines should not cross each other.

T **(F)**   **8.** Extension lines should not cross each other.

(T) F   **9.** Dimension lines should not be located within the view of an object.

T **(F)**   **10.** Dimensioning to hidden lines is considered good practice.

(T) F   **11.** A leader is a light continuous line that extends from a note to the feature of a part to which the note applies.

T **(F)**   **12.** A leader must be used with all general notes.

(T) F   **13.** A note that applies to an individual feature on a sketch is called a local note.

T **(F)**   **14.** A leader for a hole drill size should terminate at the center of the hole.

T **(F)**   **15.** The letter f shown on a drawing indicates that the surface is to be finished.

(T) F   **16.** Dimensions should be given to finished surfaces.

T **(F)**   **17.** Finish marks are placed in views that show the true shape of the surface.

T **(F)**   **18.** In the decimal inch system of dimensioning, a zero is placed before a value less than one.

(T) F   **19.** In the metric system of dimensioning, a zero is placed before a value less than one.

(T) F   **20.** An example of a general note would be: CRACKLE FINISH ALL SURFACES.

T **(F)**   **21.** A vertical dimension is a size dimension and a horizontal dimension is a location dimension.

(T) F   **22.** Dimensions should be placed in the view showing the contour shape of the part.

T **(F)**   **23.** Dimensions for the height of a feature should be placed between the front and top view.

(T) F   **24.** Dimensions are classified as size and location.

(T) F   **25.** The proper place to specify the diameter of a 12-mm drilled hole is on the view showing the hole as a circle.

T **(F)**   **26.** It is not necessary to group related dimensions.

T F   **27.** The least important dimension in a series should be omitted.

T **F**   **28.** A diameter of a cylinder is usually dimensioned in the circular view.

T **F**   **29.** The lease accurate method of dimensioning an angle is by coordinates at its end points.

T **F**   **30.** An arc is dimensioned by its diameter.

T F   **31.** Normally, an overall dimension is not required with a rounded end part.

T **F**   **32.** Countersunk holes are used for socket head screws.

T F   **33.** A hole is dimensioned in the circular view with its depth.

T **F**   **34.** If accuracy is critical for the location of holes on a bolt circle, the angle from the center in degrees should be given.

T F   **35.** The aligned system of dimensioning is no longer recommended.

36. Extension and dimension lines are constructed with a:
  a. 2B pencil.
  b. HB pencil.
  c. 3H pencil.
  d. 6H pencil.

37. The proportion, length to width, of an arrowhead is:
  a. 1:1.
  b. 2:1.
  c. 3:1.
  d. 4.1.
  e. 3:2.

38. Dimension lines should be spaced:
  a. 10 mm from the view and 10 mm from each other.
  b. 12 mm from the view and 6 mm from each other.
  c. 12 mm from the view and 12 mm from each other.
  d. 10 mm from the view and 6 mm from each other.

39. In the unidirectional system of dimensioning, dimension figures read from:
  a. the bottom and left side of the sheet.
  b. the bottom only.
  c. the bottom and right side of the sheet.
  d. the right side only.

40. The callout order for a counterbored hole is:
  a. hole size, counterbore size, counterbore depth.
  b. counterbore depth, counterbore size, hole size.
  c. hole size, counterbore depth, counterbore size.
  d. None of the above.

41. A spot facing operation is used to:
    a. provide a bearing surface.
    b. remove a burr.
    c. Neither of the above.
    d. Both (a) and (b).

42. Leaders for local notes that pertain to a hole should extend:
    a. to the center of the hole.
    b. to the edge of the hole.
    c. Neither of the above.
    d. Both (a) and (b).

43. Good dimension practice is based on standards given by the:
    a. American Institute of Industrial Engineers.
    b. American National Standards Institute.
    c. Dimensioning Council of America.
    d. None of the above.

44. Operating personnel should not have to:
    a. add or subtract dimensions.
    b. multiply or divide dimensions.
    c. scale dimensions.
    d. All of the above.

45. When applying dimensions, one should remember:
    a. reliability, cleanliness, completeness, readability.
    b. accuracy, clearness, completeness, readability.
    c. reliability, clearness, nomenclature, readability.
    d. accuracy, cleanliness, nomenclature, readability.

46. The letter R refers to:
    a. radians.
    b. a revision.
    c. the radius of an arc.
    d. the diameter of a circle.

47. Which of the following symbols would specify a spotface?
    a. ⊤.
    b. ⊔.
    c. ✓.
    d. spf.

48. Notes that do not require a leader are called:
    a. general notes.
    b. local notes.
    c. drawing notes.
    d. machining notes.
    e. None of the above.

49. Which of the following symbols specifies a depth?
    a. ✓.
    b. S.
    c. T.
    d. DEPTH.

50. The symbol × is read as:
    a. times.
    b. by.
    c. Neither of the above.
    d. Both (a) and (b).

Circle the correct answer.

T **(F)** **1.** Parts can be made exactly alike.

(T) F **2.** Variation is caused by machines, material, the environment, and operating personnel.

T **(F)** **3.** A stapler would have tighter limits than a watch.

T **(F)** **4.** The tighter the limits are, the cheaper the manufacturing process.

(T) F **5.** Limits are the extreme dimensions of a part feature.

(T) F **6.** Limit dimensioning and plus and minus tolerancing are two forms of direct dimensional tolerancing.

T **(F)** **7.** Different processes have the same degree of accuracy.

T **(F)** **8.** The "regardless of feature size" symbol is #.

T **(F)** **9.** The least material condition of a hole is the smallest hole.

(T) F **10.** The maximum material condition of a shaft is the largest shaft.

T **(F)** **11.** Selective assembly matches large holes with small shafts and small holes with large shafts.

12. The exact theoretical dimension without tolerance is the:
   a. allowance.
   b. maximum material condition.
   (c.) basic dimension.
   d. datum.

13. The smallest shaft or the largest hole is the:
   a. regardless of feature size.
   b. maximum material condition.
   c. basic size.
   (d.) least material condition.

14. Which of the following is *not* a basic way of expressing tolerance?
   a. Direct dimensional tolerancing.
   (b.) Accumulation.
   (c.) Notes.
   d. Geometric tolerancing.

15. A rectangle with symbols, letters, and/or numbers is the technique for expressing:
   a. basic dimension.
   b. datum.
   (c.) feature control frame.
   (d.) All of the above.

16. The rule that states that where a tolerance of position is specified, RFS, LMC, or MMC must also be specified with respect to the tolerance, the datum, or both.
   a. Rule 1.
   (b.) Rule 2.
   c. Rule 3.
   d. None of the above.

17. Define tolerance. _____

_____

18. Define feature and feature of size. _____

_____

_____

19. Sketch the symbols for the following:

a. Perpendicularity ___⊥___　　**b.** Total runout ___⌁⌁___

c. Cylindricity ___⌭___　　　　**d.** Position ___⌖___

e. Straightness ___—___　　　　**f.** Circularity ___○___

g. Symmetry ___⌖___　　　　　**h.** Flatness ___▱___

i. Profile of a line ___⌒___　　**j.** Concentricity ___◎___

k. Circular runout ___↗___　　**l.** Profile of a surface ___⌓___

m. Parallelism ___∥___　　　　**n.** Angularity ___∠___

20. Sketch the symbols for the following:

a. MMC ___Ⓜ___　　　　　　　**b.** RFS ___Ⓢ___

c. Diameter ___⌀___　　　　　　**d.** LMC ___Ⓛ___

e. 35 mm basic dimension ___35___　　**f.** Datum P ___P___

g. A straightness tolerance of 0.2 ___—|0.2___

Answer the questions.

1. Indicate the method of specifying the tolerance.

    Answer: _____   Answer: _____
    50.1 / 49.7   49.7 +0.4 / 0

    Answer: _____   Answer: _____
    50 +0.1 / −0.3   φ 49.7 −50.1

2. Name the three techniques of dimensioning that are associated with tolerance accumulation.

    Answer: _____   Answer: _____

    Answer: _____

3. Determine the tolerance accumulation between surface B and E for the three situations below. Also specify the method.

    Answer: _____

    Answer: _____

Answer: _____

4. Based on rule 1, determine the variation in form for a shaft with dimensions of Ø 15.3-15.5. Express your answer in the same way as the text.

5. Based on rule 1, determine the variation in form for a hole with dimensions of Ø 18.8-19.1. Express your answer in the same way as the text.

Circle the correct answer.

T   F   **1.** Datum features are associated with process or inspection equipment.

T   F   **2.** The sequence of listing datums in the feature control frame is primary, secondary, and tertiary.

T   F   **3.** For cylindrical parts, the datum is represented by two perpendicular planes that intersect at the axis of the cylinder.

T   F   **4.** The datum symbol for a cylindrical part is placed on both of the centerlines.

T   F   **5.** A datum surface has a size tolerance.

T   F   **6.** Datums of size require a material condition modifier.

T   F   **7.** A chuck on a lathe is a good example of an RFS.

T   F   **8.** Datum targets are designated points, lines, or areas of contact.

9. Three mutually perpendicular planes are called the:
  a. primary datum feature.
  b. tolerance of perpendicularity.
  c. datum reference frame.
  d. None of the above.

10. The number of points of contact for a primary datum is:
  a. one.
  b. two.
  c. three.
  d. six.

11. The number of points of contact for a secondary datum is:
  a. one.
  b. two.
  c. three.
  d. six.

12. Which of the following are datum features?
  a. Surface.
  b. Size.
  c. Location.
  d. (a) and (b).

13. Where additional support is needed and accuracy is not critical, the datum target is:
  a. a point.
  b. a line.
  c. an area.
  d. (a) and (b).

14. Sketch datum target points for an invisible secondary datum. They are 40 mm from one edge and spaced 60 mm apart. Use the letter F to identify the pins.

15. Sketch a visible datum target line for a tertiary datum 25 mm from an edge. Use the letter G to identify the pin.

16. Sketch visible target areas (Ø 10 mm) for a primary datum for a cylindrical surface that is Ø 80 mm. Use the letter H to identify the pins.

1. Sketch the appropriate tolerance *symbol* for each definition:

    _____ is a condition of a surface of revolution where all points of the surface intersected by any plane perpendicular to a common axis or passinig through a common center are equidistant from the center.

    _____ is the condition of a surface or axis at a specified angle from a datum plane or axis.

    _____ provides control of circular elements of a surface.

    _____ is the condition in which an element of a surface or an axis is a straight line.

    _____ is a condition of a surface of revolution in which all points of the surface are equidistant from a common axis.

    _____ is the condition of a surface having all elements in one plane.

    _____ is the condition of a surface, center plane, or axis at a right angle to a datum plane or axis.

    _____ is the condition of a surface equidistant at all points from a datum plane or an axis equidistant along its length to a datum axis.

    _____ is the condition where the axis of all cross-sectional elements of a surface of revolution are common to the axis of a datum reference.

    _____ is a condition in which a feature or features are symmetrically disposed about the center plane of a datum feature.

    _____ a zone in which the center, axis, or center plane is permitted to vary from the exact location, which is referred to as its true position.

    _____ provides combination control of all surface elements.

T   F   2. Straightness can only be applied to line elements in one direction.

T   F   3. Where a cylindrical feature is being specified, the diameter symbol is not necessary.

T   F   4. Where straightness is specified for an axis, the feature control frame is placed with the feature size dimension.

T   F   5. Tolerances of form require the specification of a datum.

T   F   6. Circularity is a combination of cylindricity and straightness.

T   F   7. Profile tolerancing can be applied with or without a datum.

T   F   8. Profile of a line is applied to individual line elements.

T   F   **9.** Tolerances of orientation require a datum.

T   F   **10.** Material condition modifiers are not used with tolerances of orientation.

T   F   **11.** Tolerances of location deal with features and features of size.

T   F   **12.** Tolerances of location do not require a datum.

T   F   **13.** Tolerances of location require a material condition modifier.

T   F   **14.** Square tolerance zones are not recommended because they are difficult to inspect for conformance.

T   F   **15.** The symbol PTZ stands for projected tolerance zone.

T   F   **16.** Where positional tolerance is applied at MMC for a surface, the theoretical boundary is the hole MMC minus the positional tolerance.

T   F   **17.** Where RFS is used for positional tolerance, it imposes a looser control.

T   F   **18.** Concentricity specifications should be used because verification is not difficult.

T   F   **19.** FIM stands for full indicator measurement.

T   F   **20.** Runout is most frequently used on circular objects.

21. Sketch the meaning of | ⟋ | 0.4 |

22. Given | — | 0.4 | for a feature of size (shaft) where Ø 15.3–15.7, determine the virtual condition.

1. Distinguish between ⓞ and ♮ either graphically or verbally.

2. Distinguish between a bilateral and a unilateral tolerance for ⌒.

3. Sketch the meaning of ∠ 0.5 D .

4. Sketch the meaning of // 0.2 J

5. Given   ⊥ | ⌀ 0.3 | B Ⓜ   for a feature of size (shaft) of ⌀ 32.4–32.8, determine
    the tolerance zone at MMC and LMC. What is the bonus tolerance?

6. Sketch a hole (⌀ 20 ± 0.2) and properly dimension its true position with a 0.3
    diameter tolerance zone at MMC. The hole is located 35 and 45 mm from a corner
    which is the intersection of datums A and B.

7. Given a hole (⌀ 20.3–20.8) and a positional tolerance of ⌀ 0.4 Ⓜ for the axis,
    determine the positional tolerance at MMC and LMC. What is the additional toler-
    ance?

8. Distinguish between circular and total runout by means of a sketch or verbal discus-
    sion.

Circle the correct answer.

T   F   **1.** Information in parentheses indicates that it is for reference only.

T   F   **2.** Clearance fits have an air space between the mating parts.

T   F   **3.** Interference fits sometimes have material interference and sometimes an air space.

T   F   **4.** Transition fits are used for location.

T   F   **5.** The hole basis system uses the smallest hole as the basic size.

T   F   **6.** The shaft basis system uses the smallest shaft as the basic size.

T   F   **7.** The lower deviation for the shaft in the shaft basis system is 0.

T   F   **8.** The lower deviation for the hole in the hole basis system is 0.

T   F   **9.** There is an air space in a transition fit when both parts are at LMC.

T   F   **10.** The fundamental deviation of the shaft is a lowercase letter.

11. Define *basic size*. _____

_____

_____ .

12. Define *upper deviation*. _____

_____

_____ .

13. Define *lower deviation*. _____

_____

_____ .

14. Define *fundamental deviation*. _____

_____

_____ .

15. Define *international tolerance*. _____

_____

_____ .

16. Define *tolerance zone.* ——————————————————

——————————————————————————————————

——————————————————————————————————

17. Using a close-running fit for two mating parts and the shaft basis system, determine the hole and shaft tolerance symbols, basic size, and limit dimensions. The design calculation for the diameter is 7 mm. Also calculate the tightest and loosest fit.

18. Using a locational interference fit for two mating parts and the hole basis system, determine the hole and shaft tolerance symbols, basic size, and limit dimensions. The design calculation for the diameter is 12 mm. Also calculate the tightest and loosest fit.

1. Define *roughness*. _____

_____

_____

2. Define *surface texture*. _____

_____

_____

3. Define *roughness height*. _____

_____

_____

4. Define *roughness width*. _____

_____

_____

5. Define *roughness-width cutoff*. _____

_____

_____

6. Define *lay*. _____

_____

_____

7. Define *waviness*. _____

_____

_____

8. Define *waviness height*. _____

_____

_____

9. Define *waviness width.* _____

_____

_____

10. Express a roughness width of 0.1, roughness width cutoff of 0.3, waviness height of 0.008, roughness height of 2.3, parallel lay, and waviness width of 4 in the symbol below.

11. Which metal removal processes would be recommended for a roughness average of 0.80 μm?

12. Which metal removal processes would be recommended for a roughness average of 0.20 μm?

Circle the correct answer.

T **(F)** 1. Screw threads are classed as permanent fasteners.

(T) F 2. Threaded fasteners are manufactured in standard types and sizes.

(T) F 3. Threads may be either internal or external.

T **(F)** 4. Thread forms used in U.S. industry are based on approved European standards.

(T) F 5. Screw threads are used for holding parts together, adjusting parts, and transmitting power.

T **(F)** 6. Threads can only be manufactured on a lathe.

(T) F 7. Depending on the application, a thread may be either left hand or right hand.

T **(F)** 8. Most threads in use today are right hand; therefore, the letters **RH** are placed at the end of the thread, not on the drawing.

(T) F 9. Lead is the distance a screw thread advances in one turn measured parallel to the axis.

T **(F)** 10. In a double thread, the pitch is equal to twice the lead.

(T) F 11. Threads may be represented on drawings using one of three conventional methods: detailed, schematic, and simplified.

(T) F 12. To save drafting time, the simplified convention is used.

T **(F)** 13. The unified system has two graded pitch series, and the metric system has three graded pitch series.

T **(F)** 14. The unified system has a primary and a secondary choice for the nominal size, while the metric has these two plus a third.

T **(F)** 15. In a thread note, UNF stands for united national fit.

T **(F)** 16. In the unified system, a class 3 fit has a greater allowance between mating parts and has closer tolerances than either class 1 or 2 fit.

T **(F)** 17. A 12 UNC-2 thread would provide finer adjustments than a 20 UNF-3 thread.

(T) F 18. In the metric system, a tolerance grade of 8 is a coarser fit than a lower number.

(T) F 19. In the metric system, the allowance is specified by a letter, and in addition, capital letters stand for an internal thread and lowercase letter for an external thread.

(T) F 20. The letter "e" can stand for a large allowance on a hole.

(T) F 21. Screw threads are identified on a sketch by a standard note with a leader and an arrow pointing to the area of the part to be threaded.

(T) F 22. A right-hand thread advances into the hole when turned clockwise.

T **(F)** 23. In the threaded note .25 20 NC-1, the .25 is the length of the threaded portion of the bolt.

T F **24.** The purpose of thread relief is to relieve the material so that the threading die may pass over the relief and make the threads complete to the end of the piece, thereby allowing the nut to be screwed tightly against the shoulder.

T F **25.** Finished nuts and bolts have a washer face.

T F **26.** Cap screws join two threaded parts together.

T **F** **27.** Machine screws are used for joining thick materials.

T F **28.** Setscrews are used to prevent motion between two parts.

T F **29.** One example of a permanent fastener is the rivet.

T F **30.** Size of the rivet is designated by body diameter and usage.

T F **31.** Welding processes are classified into three main types: arc, gas, and resistance.

T F **32.** Brazing is the process of joining metals with a nonferrous filler metal that has a melting point lower than the joining metals.

T F **33.** Keys are a type of fastener that are used to prevent rotation of gears, pulleys, and collars on rotating shafts.

T F **34.** A type of locking device that is used to prevent a nut from loosening on a bolt is a jam nut.

T F **35.** Chemical fastening is a process used by the aircraft industry.

EXERCISE 9-2

Name: _____

Fill in the spaces below.

1. In the space provided, sketch the following screw thread representations:

    a. Simplified external   **b.** Simplified internal

    c. Schematic external   **d.** Schematic internal

2. Sketch an external thread giving the crest, root, pitch, major diameter, minor diameter, pitch diameter, depth, and angle.

1. Using the unified and metric tables, fill in the missing information:

	NOMINAL SIZE (BASIC MAJOR DIAMETER)	PITCH	METRIC/UNIFIED	SERIES
a.	1.6	0.2	m	Constant
b.	4 (0.1120)	1/48 = 0.02	U	FINE UNF GRADED
c.	5	0.8	m	Coarse
d.	7/16 .4375	14	U	UNC
e.	5.5	0.5	Metric	COUSTANT

2. Fill in the spaces below.
   a. Interpret the following thread specification: .750-10 UNC-2A.

   .750 ___NOMINAL DIAMETER_____

   10 ___PITCH (THREAD/INCH) COARSE___

   UNC ___UNITED COARSE_____

   2 ___CLASS OF FIT - NORMAL_____

   A ___EXTERNAL_____

   b. Interpret the following specification for an internal thread: .3125-24 UNF-2B-LH.

   .3125 ___NOMINAL DIAM_____

   24 ___THREAD/INCH_____

   UNF ___UNITED FINE_____

   2 ___NORMAL_____

   B ___INTERNAL_____

   LH ___LEFT HAND THREAD_____

c. Interpret the following specification for a metric thread: M10 × 1.5-6H.

M _____ METRIC _____

10 _____ SIZE IN MM (NOMINAL DIA) _____

1.5 _____ PITCH _____

6 _____ NORMAL FIT TOLERANCE COARSE _____

H _____ NO ALLOWANCE _____

d. Interpret the following specification for a metric thread: M12 × 1-6g L.

M _____ METRIC _____

12 _____ SIZE IN MM _____

1 _____ PITCH _____

6 _____ NORMAL FIT _____

g _____ SMALL ALLOWANCE _____

L _____ LONG _____

3. Provide the information for the four figures.

1.2 - 5 ACME - 2G DBL   M5 x 0.5 - 6h LH

M12 x 1.75 - 5g   1.1250 - 18 UNEF - 3A

(a)   (b)   (c)   (d)

		(a)	(b)	(c)	(d)
1.	Number of threads per inch	NA	5	NA	18
2.	RH or LH thread	RH	RH	LH	RH
3.	Thread form used	METRIC	ACME	METRIC	UNIFIED
4.	Tolerance (class of fit)	5g	2G	6h	3A
5.	Single or which multiple thread	S	DBL	S	S
6.	Pitch of thread	1.75	1/5 "	0.5	1/18 "
7.	Lead of thread	1.75	2/5 "	0.5	1/18 "
8.	Major diameter	12 mm	1.2	5 mm	1.1250
9.	Internal or external thread	E	I	E	E
10.	Method of representation	SIMPLIFIED	DETAILED	SCHEMATIC	DETAIL

EXERCISE 10-1

Name: _____

Circle the correct answer.

T F   **1.** The term "working drawing" includes both detail and assembly drawings.

T F   **2.** A detail drawing is a completely dimensional shop drawing made to scale.

T F   **3.** A detail drawing shows one part only.

T F   **4.** An assembly drawing must show at least three parts.

T F   **5.** A detail drawing sketch is a completely dimensional sketch.

T F   **6.** The detail drawings would be made before the design layout is made.

T F   **7.** A subassembly drawing is made showing only one part.

T F   **8.** A detail drawing and a detail view are the same.

T F   **9.** Reading a drawing involves visualization and interpretation.

T F   **10.** Title blocks provide the basic information about the part or assembly.

T F   **11.** One reason for a revision to a drawing is improved design.

T F   **12.** The bill of materials contains the list of the parts and/or material used on or required by an assembly drawing.

T F   **13.** The first step in reading a drawing is to analyze the part or subassembly or assembly (whichever the case).

T F   **14.** Before a drawing is said to be finished, it must be checked for completeness, accuracy, and clarity. The checker should then sign or initial the drawing in the appropriate place in the title block.

T F   **15.** The revision record shows what was before the change was made.

T F   **16.** The bill of materials shows manufactured parts as well as purchased parts.

T F   **17.** When analyzing the part, concentrate on the front view only, as the other views were made to take up time of the drafter.

T F   **18.** The last step in analyzing the part is to study one feature at a time and begin to picture in your mind the shape of the real object.

T F   **19.** A detail drawing illustrates a single component or part, showing a complete and exact description of its shape, dimensions, and features.

T F   **20.** The bill of materials is found only on detail drawings.

Answer the questions.

DRAWN: *DHB*	11/13	NORTH AMERICAN UNIVERSITY
CHECK: *gg*	11/15	COLLEGE OF ENGR. AND TECH.
DESIGN:		ANYTOWN, USA
APPROVED: *BO'H*	11/20	STEM
APPROVED:		DWG. SIZE A4 — 45438-2
MATERIAL: 1.25 RD BR ROD		SCALE: 1"=1" WT: .05# SHEET: 1 OF 1

1. What is the name of the part? _STEM_

2. Give the drawing number. _45438-2_

3. How many sheets are in the set? _1_

4. What is the weight of the part? _.05 lb_

5. To what scale was the drawing made? _1"=1"_

6. What material is specified? _1.25 Round BR ROD_

7. Who made the drawing? When? _DHB 11/13_

8. How many people have initialed the drawing? _3_

Answer the questions.

DRAWN: _jan_	2/20	NORTH AMERICAN UNIVERSITY
CHECK: _gg_	2/21	COLLEGE OF ENGR. AND TECH.
DESIGN:		ANYTOWN, USA
APPROVED: _BH_	2/24	VALVE BODY
APPROVED: _BB_	2/25	DWG. SIZE A4 · 45438-1
MATERIAL: FORGING 11894		SCALE: 1 : 2 · WT: 15.3 Kg · SHEET: 1 OF 2

1. Give the name of the part.

   VALVE BODY

2. What is the drawing number?

   45438 -1

3. How many sheets are in the set?

   2

4. Give the scale of the drawing.

   1 : 2

5. What size of drawing was made?

   A4 → 8.5 × 11"

6. How many people have initialed, indicating their approval of the drawing, besides the drafter?

   2

7. What material is specified?

   FORGING 11894

8. Who approved the drawing? When?

   BH / 2/24   BB / 2/25

9. What is the weight of the part?

   15.3 kg

Answer the questions.

		REVISIONS		
ZONE	REV.	DESCRIPTION	DATE	APPROVED
C5	1	WAS 42.6 DIA.	07/10/88	*B. Boss*
B4	2	WAS M24 X 3 X 50	09/26/88	*B. Boss*
D4	3	FINISH ALL OVER ADDED	11/30/88	*T L*
A2	4	SECTION B-B ADDED	01/03/89	*B. O'H*

1. How many changes are listed? _____4_____

2. What was the change in revision 1? __CHANGE FROM 42.6 DIAM__

3. In what zone is revision 2 listed? ___B4___

4. In revision 3, what was added? ___FINISH___

5. What is the date of the first revision? __7/10/88__

		REVISIONS		
ZONE	REV.	DESCRIPTION	DATE	APPROVED
	A	3 X 45° CHAMFER ADDED	06/12/88	*B. Boss*
	B	WAS REAM AFTER DRILLING	09/15/88	*B. Boss*
	C	NOTE ADDED	02/23/89	*B. O'H*

1. How many changes are listed? _____3_____

2. What was the change in revision B? __WAS REAM AFTER DRILLING__

3. What was added in revision C? ___NOTE___

4. What was the size of the chamfer added? __3 X 45°__

5. What is the date of the last revision? __2/23/89__

EXERCISE 10-3   Name: _____

Answer the questions.

7	1	CAP SC.	1112 ST.	M10 X 1.5 X 50
6	2	STEM	STN. STL.	TYPE 316
5	2	HANDLE	BRASS	Ø 12 ROD
4	1	BUSHING	BRONZE	OILITE #1612
3	1	O RING	RUBBER	BUNA N SYN
2	3	WASHER	1020	CHROME PLATE
1	1	BASE	C.I.	
ITEM	QTY.	NAME	MATERIAL	SPECIFICATIONS
			BILL OF MATERIALS	

1. What is the block called that lists the materials?   _BMATERIAL MATERIALS_

2. How many different parts are listed?   _7_

3. Give:
   a. the name of item 3   _O - RING_

   b. the material   _RUBBER_

   c. the specifications   _BUNA N SYN_

4. For the stem give:
   a. the material   _STN. STL._

   b. the specification   _TYPE 316_

5. For item 2 give:
   a. the quantity required   _3_

   b. the part name   _WASHER_

   c. the material   _1020 STEEL_

   d. the material finish   _CHROME PLATE_

6. For the handle give:
   a. the item
   _____ 5 _____ HANDLE _____

   b. the specification
   _____ Ø12 ROD _____

   c. the material
   _____ BRASS _____

   d. the quantity
   _____ 2 _____

Answer the questions.

ITEM	QTY.	NAME	MATERIAL	SPECIFICATIONS
5	2	NUT	1112 ST.	M12 X 2 SQ.
4	2	WASHER	1112 ST.	M12
3	1	BUTTON	BRONZE	Ø 12
2	6	SPRING	MUS. WIRE	1 MM
1	1	VALVE	C.I.	PATTERN NUMBER 352-K

BILL OF MATERIALS

1. How many different parts are listed?
_____ 5 _____

2. Give the name of item 5.
_____ NUT _____

3. For the valve give:
   a. the specification
   _____ PATTERN # 352-K _____

   b. the material
   _____ C. I. _____

4. For the spring give:
   a. the quantity required
   _____ 6 _____

   b. the material
   _____ MUS. WIRE _____

   c. the specification
   _____ 1 MM _____

5. What material is required for the button?
_____ BRONZE _____

6. For item 5, what is:
   a. the part name
   _____ NUT _____

   b. the specification
   _____ M12X2 SQ _____

Answer the questions.

TOOL STEEL	245 – A – 12	16	SHOULDER PIN	
MATERIAL	ORDER NO.	QTY		

1. Name the view represented.   _FRONT_

2. What is the shape of the shoulder pin?   _114 X 58 (CYLINDRICAL)_

3. How many outside diameters are shown?   _4_

4. What is the largest diameter?   _58_

5. What diameter is the smallest hole?   _12_

6. What is the overall length of the pin?   _114_

7. How deep is the Ø18 hole?   _27_

8. How wide is the Ø44 portion?   _26_

9. What letters represent object lines?   _A, C, & F_

10. What kinds of lines are **B** and **D**?   _HIDDEN_

EXERCISE 10-4 (Continued)

Name: _____

11. What letter represents the centerline? _____ *E*

12. What does the centerline indicate about the holes and outside diameters? ___ *CONCENTRIC — ALL HAVE SAME CENTERLINE*

13. Give the width of the Ø58 portion. _____ *98*

14. State the order number of the part. _____ *245-A-12*

15. What is the material specified for the pins? _____ *TOOL STEEL*

Answer the questions.

M20 X 1.5 - 6g

3

Ø 16

Ø 25.4 $^{+0}_{-0.1}$

M24 X 3 X 30

Ø 6 ±0.05 THRU

430

450

475

30° CHAMFER TO THREAD DEPTH - 2X

UNLESS OTHERWISE SPECIFIED
ALL DIMENSIONS ±0.1

METRIC

1. Give the finished length including the tolerance permitted. _____ *475 ± 0.1 mm*

2. What size of hole is specified? _____ *6 ± 0.05*

3. Interpret the thread specification on the right end. _____ *24 mm / PITCH 3 / LENGTH*

4. Give the dimensions for the thread relief. _____ *3 m x Ø16*

5. How long is the thread on the left? _____ *25 22 ± 0.1*

6. Give the limit dimensions for the shaft diameter. _____ *+0 / -0.1   25.4 / 25.3*

7. Why are the break lines used? _____ *OBJECT IS TOO LONG TO SHOW ON PAGE*

8. What is the chamfer specification on the left end? _____ *30° TO THREAD DEPTH (TWICE)*

9. Explain the tolerance and allowance specification on the M20 thread. _____ *NORMAL FIT - EXTERNAL*

298

Answer the questions.

1. What is the name of the part? _____ SPINDLE SHAFT

2. What is the material specification? _____ SAE 2335

3. How many parts are required? _____ 42

4. What type of heat treatment is called for? _____ HARDEN & TEMPER

5. What is the order number? _____ L224-1

6. What is the hole depth? _____ 9 mm

7. Find the following lengths: **A, B, C, D.**

A _____ 159 ± 0.1 _____

B _____ 16 _____

C _____ 13 _____

D _____ 63 _____

8. What is the largest diameter? _____ 58 _____

9. What is the largest thread diameter? _____ 30 _____

   Smallest thread diameter? _____ 16 _____

10. What is the tolerance of the Ø28 dimension? _____ +0 -0.05 _____

11. What is the length of the Ø25 portion of the shaft? _____ 25 ± 0.1 _____

12. What is the length of the M30 thread? _____ 15 _____

    The M16 thread? _____ 22 _____

13. Which thread is a coarse series? _____ M30 x 2 _____

14. What is the final operation on the diameters? _____ GRIND AFTER HARDNING _____

15. What does the (38) dimension mean? _____ REFERENCE _____

Answer the questions.

1. What is the name of the part? _____ ROCKER BASE _____

2. What type of material is used? _____ CAST STEEL _____

3. Is this part heat-treated? What type of treatment is specified? _____ NORMALIZE AFTER ROUGH MACH. _____

4. How many general notes are there? _____ 2 _____

5. How many local notes? _____ 7 _____

6. What are the limits on dimensions for the following:

     a. Fractional _____ ± 1/64" _____

     b. Decimal _____ ± .001" _____

     c. Angular _____ ± 15' _____

7. How many holes are reamed? _____ 3 _____

8. What is the length of the $1\frac{1}{2}$-16UN-3LH thread? _____ 3/4" _____

9. What radius is specified for fillets and rounds? _____ 1/8 _____

10. What are the lengths for the following: A, B, C, D, E.

A _____ 3.374 – 3.375 _____

B _____ 4.446 – 4.45 _____

C _____ 3.25 ± .002/.000 _____

D _____ 4.8125   1.5625 _____

E _____ 0.625   .8125 _____

11. What are the limit dimensions for the center line distance between the $\frac{55}{64}$ drill $\frac{7}{8}$ ream hole and $\frac{3}{4}$- 16UNF - 2LH? _____ 1.248 – 1.250 _____

12. Describe the following thread callout:

$1\frac{1}{2}$ _____ DIAM _____

16 _____ THREAD / INCH _____

UN _____

3LH _____

13. Describe the following local note callout:

No. 10 _____

24 _____

NC _____

$\frac{7}{16}$ ————————————————————

4 ————————————————————

14. Express the centerline distance between the $\frac{55}{64}$ drill, $\frac{7}{8}$ ream hole and $\frac{31}{64}$ drill, $\frac{1}{2}$ ream 2 holes as a plus-and-minus tolerance dimension.

————————————————————

15. What is the length of the unthreaded part of the $1\frac{1}{2}$- 16UN - 3LH thread?

————————————————————

Answer the questions. See the print on the following page.

1. How many surfaces are to be finished? 6 7

2. Except where noted otherwise, what is the tolerance on **A** and **B**?

   A ±0.2   50 ±0 −0.05

   B ±0.2

3. What is the tolerance on the Ø12 holes? +8.5

4. What are the limit dimensions for the 40 dimension shown on the side view? + 0.10 − 0.05

5. What are the limit dimensions for the 25 dimension shown on the side view? ±0.2

6. What is the maximum permissible distance between the centers of the Ø8 holes? Ø0.4

7. Express the Ø12 hole as a plus-and-minus tolerance dimension. Ø12 + .05 − .00

8. What is the maximum tolerance placed on dimension **H**? _____

9. Calculate distances **A** to **N**:

   A 50

   B 15

   C 2

   D 44

   E 28

   F 4

   G 11

   H 75

   J 6

   K 13

   L 0.4

NOTE: EXCEPT WHERE NOTED --
TOLERANCES +/- 0.2
ROUNDS AND FILLETS 3 R

METRIC

M _____

N _____

10. What are the limit dimensions for the 6 dimension shown on the top view? _____

11. Locate surfaces **4** on the top view. _____ 8 ∮ 10 ✓ _____

12. How many surface **5**'s are there? _____ 4 _____

13. Locate line **3** in the top view. _____ ✓ _____

14. Locate line **6** in the side view. _____ ✓ _____

15. What surfaces in the front view indicate line **4** in the side view? _____ 7, 11 _____

EXERCISE 11-1

Name: _____

Circle the correct answer.

T **(F)** 1. CAD/CAM is the use of computers to aid the design and material acquisition of a product.

(T) F 2. The first step is usually to communicate ideas by means of a technical sketch.

(T) F 3. The ability to work with a computer-controlled drawing system is essential.

T **(F)** 4. CAD/CAM are independent activities.

T **(F)** 5. CAD/CAM is an outgrowth of the development of microcomputers.

T **(F)** 6. By the mid-1970s, CAD systems had become quite commonplace in manufacturing.

(T) **(F)** 7. Minicomputers are smaller versions of large mainframe computers.

(T) F 8. The CPU executes instructions.

(T) F 9. Menus give easy access to specific instructions or drawing routines.

T **(F)** 10. The only way to move the cursor is by a mouse.

T **(F)** 11. With a drum plotter, the pen moves both vertically and horizontally.

(T) F 12. Dot-matrix printers are relatively inexpensive, but line quality is not as good as a plotter.

T **(F)** 13. CAD software is divided into the operating system and the graphics system.

(T) F 14. One major advantage of CAD is that the link with manufacturing is easily effected.

(T) F 15. Menus are hierarchically structured.

(T) F 16. The primary input methods are keyboard, digitizer/puck or light pen, and mouse.

T **(F)** 17. Menus are only displayed on the screen.

T **(F)** 18. Mirroring is used with objects that are not symmetrical.

(T) F 19. The copy function saves the drafter considerable time when a number of objects is repeated.

T **(F)** 20. One of the least desirable enhancements of a CAD system is the use of spacial coordinates.

21. CAD is defined as:
    (a.) computer-aided design.
    b. computer-aided drafting.
    c. computer-aided design and drafting.
    (d.) All of the above.

22. Three types of magnetic storage media are:
    (a.) cartridge, floppy disk, hard disk.
    b. cartridge, floppy disk, paper tape.
    c. cartridge, hard disk, paper tape.
    (d.) hard disk, paper tape, floppy disk.

23. A small, low-cost computer is called a:
    a. maincomputer.
    b. microcomputer.
    c. minicomputer.
    d. small computer.

24. Which of the following is *not* a component of a computer system?
    a. Storage.
    b. Production.
    c. Input.
    d. Output.

25. Which of the following is *not* a function of a workstation?
    a. Interface with the CPU.
    b. Provide digital descriptions of the graphic image.
    c. Facilitate communication between the user and the system.
    d. All of the above.

26. Which of the following is an output device?
    a. keyboard.   **c.** printer.
    b. digitizer/light pen.   **d.** mouse.

27. The best device for inputting textual material is a:
    a. keyboard.   **c.** mouse.
    b. printer.   **d.** digitizer/puck.

28. A particular location on the screen is given by a cross-hair called a:
    a. pointer.   **c.** cursor.
    b. bracket.   **d.** cross-hair.

29. Which of the following are commercially available software systems?
    a. CADKEY.   **c.** AUTOCAD.
    b. VERSACAD.   **d.** All of the above.

30. The zoom function enables the system to:
    a. increase the image speed.
    b. make the image larger.
    c. speed up the plotter.
    d. None of the above.

31. The process of turning a part displayed on the CRT through a selected angle about an axis is called:
    a. rotating.   **c.** moving.
    b. turning.   **d.** (a) and (c).

32. Which of the following are common CAD system enhancements?
    a. Crosshatching.
    b. Consistency.
    c. Automatic dimensioning.
    d. (a) and (c).

33. As compared to instrument drawing, which of the following is *not* an advantage of CAD?
    a. Faster.   **d.** Neater.
    b. More accurate.   **e.** None of the above.
    c. More consistent.

INDEX